Strategic Studies Institute
and
U.S. Army War College Press

TOWARD A UNIFIED MILITARY RESPONSE: HURRICANE SANDY AND THE DUAL STATUS COMMANDER

Ryan Burke
Sue McNeil

April 2015

The United States Army War College

The United States Army War College educates and develops leaders for service at the strategic level while advancing knowledge in the global application of Landpower.

The purpose of the United States Army War College is to produce graduates who are skilled critical thinkers and complex problem solvers. Concurrently, it is our duty to the U.S. Army to also act as a "think factory" for commanders and civilian leaders at the strategic level worldwide and routinely engage in discourse and debate concerning the role of ground forces in achieving national security objectives.

The Strategic Studies Institute publishes national security and strategic research and analysis to influence policy debate and bridge the gap between military and academia.

The Center for Strategic Leadership and Development contributes to the education of world class senior leaders, develops expert knowledge, and provides solutions to strategic Army issues affecting the national security community.

The Peacekeeping and Stability Operations Institute provides subject matter expertise, technical review, and writing expertise to agencies that develop stability operations concepts and doctrines.

The Senior Leader Development and Resiliency program supports the United States Army War College's lines of effort to educate strategic leaders and provide well-being education and support by developing self-awareness through leader feedback and leader resiliency.

The School of Strategic Landpower develops strategic leaders by providing a strong foundation of wisdom grounded in mastery of the profession of arms, and by serving as a crucible for educating future leaders in the analysis, evaluation, and refinement of professional expertise in war, strategy, operations, national security, resource management, and responsible command.

The U.S. Army Heritage and Education Center acquires, conserves, and exhibits historical materials for use to support the U.S. Army, educate an international audience, and honor Soldiers—past and present.

STRATEGIC STUDIES INSTITUTE

The Strategic Studies Institute (SSI) is part of the U.S. Army War College and is the strategic-level study agent for issues related to national security and military strategy with emphasis on geostrategic analysis.

The mission of SSI is to use independent analysis to conduct strategic studies that develop policy recommendations on:

- Strategy, planning, and policy for joint and combined employment of military forces;
- Regional strategic appraisals;
- The nature of land warfare;
- Matters affecting the Army's future;
- The concepts, philosophy, and theory of strategy; and,
- Other issues of importance to the leadership of the Army.

Studies produced by civilian and military analysts concern topics having strategic implications for the Army, the Department of Defense, and the larger national security community.

In addition to its studies, SSI publishes special reports on topics of special or immediate interest. These include edited proceedings of conferences and topically oriented roundtables, expanded trip reports, and quick-reaction responses to senior Army leaders.

The Institute provides a valuable analytical capability within the Army to address strategic and other issues in support of Army participation in national security policy formulation.

*****

Comments pertaining to this report are invited and should be forwarded to: Director, Strategic Studies Institute and U.S. Army War College Press, U.S. Army War College, 47 Ashburn Drive, Carlisle, PA 17013-5010.

*****

This manuscript was funded by the U.S. Army War College External Research Associates Program. Information on this program is available on our website, *www.StrategicStudiesInstitute.army.mil*, at the Opportunities tab.

*****

*****

The Strategic Studies Institute and U.S. Army War College Press publishes a monthly email newsletter to update the national security community on the research of our analysts, recent and forthcoming publications, and upcoming conferences sponsored by the Institute. Each newsletter also provides a strategic commentary by one of our research analysts. If you are interested in receiving this newsletter, please subscribe on the SSI website at *www.StrategicStudiesInstitute.army.mil/newsletter.*

ISBN 1-58487-677-8

CONTENTS

FOREWORD

The military response to Hurricane Sandy was historically significant. For the first time, dual status commanders—military commanders authorized by law to serve in both state and federal status simultaneously—were activated in an effort to unify state and federal efforts under a single command structure during an unplanned response. The events that occurred over the 2-week response period provide a basis for lessons learned as we look to improve upon this first attempt to use this unique command construct in response to an unplanned incident.

In this monograph, Mr. Ryan Burke and Dr. Sue McNeil discuss the results of their year-long case study examining the military response to Hurricane Sandy under the dual status commander arrangement. In one of the most thorough and comprehensive analyses of the subject to date, Burke and McNeil first address the chronology of the storm and the ensuing military actions. They then assess both the successes and shortfalls with the storm response. The authors conclude the monograph with a detailed discussion of 15 recommendations, which have the potential to significantly improve future dual status commander-led civil support operations.

DOUGLAS C. LOVELACE, JR.
Director
Strategic Studies Institute and
U.S. Army War College Press

ABOUT THE AUTHORS

RYAN BURKE is currently a doctoral candidate at the University of Delaware, with a research focus on military civil support operations. Before starting his Ph.D., he was a captain and logistics officer in the U.S. Marine Corps, where he served as a landing support platoon commander, operations officer, and company commander with Combat Logistics Regiment 27, 2d Marine Logistics Group, Camp Lejeune, NC. He then served as the Deputy Marine Officer Instructor for the Philadelphia Naval ROTC Consortium at the University of Pennsylvania. After leaving the Marines, he worked as a Senior Consultant and Logistics Analyst for Booz Allen Hamilton supporting several Department of Defense projects both in the Pentagon and aboard Marine Corps Base Quantico. Mr. Burke holds a bachelor's degree in crime, law, and justice and a minor in military science from the Pennsylvania State University where he attended on a Naval Reserve Officers Training Corps scholarship, and a master of science in homeland security from Saint Joseph's University.

SUE MCNEIL is Professor of Civil and Environmental Engineering, and Public Policy and Administration at University of Delaware. She also directs the Graduate Program in Disaster Science and Management and is the former Director of the Disaster Research Center at the University of Delaware. She has extensive experience working with transportation organizations and agencies on multicriteria decisionmaking, performance measures, target setting, and asset management. Dr. McNeil has served as a member of the Executive Committee of the Transportation Research Board and a member of the Board on Infrastructure and the Constructed Environment. She chaired the Transportation Research Board's Committee on Asset Management from 2004 to 2010 and is an Emeritus Member of the committee. She is a founding Associate Editor for the American Society of Civil Engineers' *Journal of Infrastructure Systems* and became the Editor-in-Chief in July 2010. She is a registered professional engineer.

SUMMARY

U.S. military forces have played a role in supporting civil authorities in varying locations and capacities from the Whiskey Rebellion to Hurricane Sandy. In a large-scale incident response scenario requiring combined support from the National Guard and federal military, effective management and coordination continues to challenge all involved. There are issues of constitutionality, legality, policy, financial considerations, and even politics, all uniquely situated between individual states' interests and those of the Federal Government. In this context, there is a philosophical conflict between federalism and state sovereignty during military civil support missions that continues to present itself as an impediment to success. Balancing these institutionally divergent approaches to achieve a unified, efficient, coordinated, and effective military response has, and will continue to be, a strategic and political imperative. Despite the challenges, military forces are frequently involved in domestic response missions, often in a very public manner. As such, military force allocation and management have evolved into major topics of conversation among policymakers, academics, emergency managers, and military strategists alike.

Owing to these issues, State and Federal Government lawmakers adopted policy and law authorizing a single military commander, referred to as a dual status commander, to legally assume simultaneous but mutually exclusive command and control over both Title 32 and Title 10 forces during domestic operations. As a proposed solution to the notable coordination challenges plaguing domestic civil support operations, the dual status commander initiative has been used successfully during planned events since 2004. The coordinated military response to Hurricane Sandy in the fall of 2012 was the first time in U.S. history dual status commanders assumed command of both Title 10 and Title 32 forces during a no-notice/limited-notice incident. As such, this event provides a relevant and timely opportunity to study the military response to the storm and offer objective recommendations for improving future no-notice/limited-notice defense support of civil authorities (DSCA) operations under the dual status commander arrangement.

The purpose of this monograph, therefore, is to offer an objective and systematic documentation and evaluation of the military response to Hurricane Sandy as a basis for assessing the efficacy of the dual status commander arrangement for no-notice/limited-notice incidents in the homeland. To complete this effort, we employed a rigorous case study investigation emphasizing the combined state and federal response to Hurricane Sandy in the New York City metropolitan area from October 22-November 15, 2012. The research examines the events of the storm response under the command of Brigadier General Michael Swezey, the designated dual status commander for the storm response in New York. We combined interviews with Department of Defense officials, National Guard commanders, and Active Duty military officers involved in the Sandy response with extensive document and content analysis of various Sandy-specific reports to generate our findings. Through this research, we intend to present a detailed and objective analysis of the response in order to provide military and defense officials with a greater understanding of the benefits and limitations of the dual status commander arrangement during a no-notice/limited-notice civil support incident. We conclude by offering a series of recommendations likely to improve policy, procedures, and training, among other things.

CHAPTER 1

INTRODUCTION

With often-unparalleled emergency response capabilities and capacities, the U.S. military is an extraordinary asset capable of providing immediate assistance to civil authorities to "save lives, prevent human suffering, and mitigate great property damage within the United States."[1] The strategic rebalancing of defense priorities away from counterterrorism and insurgency operations and back to the homeland will see the U.S. military continue to play an important role in domestic civil support and crisis response operations. As such, it is imperative that we continue to learn from past events like Hurricane Katrina and Hurricane Sandy and strive for continuous improvement of our military's domestic response capabilities.

Hurricane Sandy was the largest and most damaging Atlantic hurricane on record and the second most costly in U.S. history, eclipsed only by Hurricane Katrina.[2] At the peak of the October-November 2012 military response to Hurricane Sandy in New York, more than 4,000 National Guard personnel, along with Active and Reserve Soldiers, Sailors, Airmen, and Marines were engaged in supporting civil authorities as part of Joint Task Force Sandy. Most of our substantive post-Sandy knowledge is found in various Department of Defense (DoD) after-action reports and lessons learned publications.[3] Beyond DoD publications and a small body of news reports, there is a dearth of knowledge to date specifically analyzing the DoD response to Hurricane Sandy available for public consumption. Further, this particular defense support of civil authorities (DSCA) event marked the historic first-time use of a dual status commander to command both Title 10 and Title 32 force components in support of a no-notice/limited-notice[4] incident. Owing to the newness of the concept and because it had not been used in an actual disaster before Sandy, our findings show a range of perspectives among DoD and state personnel: some advocating for and supporting the concept, others noting the concept's limitations and challenges. Regardless of position, it is clear that the dual status commander arrangement has several benefits and limitations when applied to a no-notice/limited-notice DSCA response environment. Our study analyzes the dual status commander-led DSCA response to Hurricane Sandy in New York. Through this lens, the study illustrates and discusses the perspectives of the dual status commander construct and offers recommendations for leveraging existing capabilities and improving those deemed insufficient.

STUDY CONTEXT

In October 2012, Hurricane Sandy came ashore along one of the most densely populated regions in the country. Even though Sandy was downgraded to tropical storm status prior to landfall, it was a massive storm that affected east coast cities from Washington, DC, to New York City. As was the case with Katrina, the storm's magnitude overwhelmed state and local responders. Requests for military support were widespread, resulting in an over-convergence of military forces inside the region within days of the storm's arrival. Again, like Katrina, National Guard forces in both State Active Duty and Title 32 status operated alongside Title 10 federal forces in support of civil authorities responding to the storm's damage. Unlike

Katrina, however, this DSCA response effort was a historical first for the U.S. military. For the first time, National Guard and federal military forces executed unplanned civil support operations under the tactical command of dual status commanders.

In addition to being the first no-notice/limited-notice operation to combine Title 32 and Title 10 forces under the command of a single dual status commander, a daunting endeavor in its own right, several additional factors further complicated the Hurricane Sandy response effort. While the storm made landfall nearly 100 miles south of New York City, the leading northern edge of the storm, widely held as the strongest part of any hurricane, directly impacted New York City, its surrounding boroughs, and parts of northern New Jersey. This ultimately resulted in a multistate incident spread across a large area of some of the most densely populated counties in the United States. But perhaps the most significant complication, as it turns out, was the storm's unprecedented timing. The 2012 presidential election was set to take place nearly a week to the day, following the storm's landfall. As we have seen with past disasters and emergencies, such events can serve as a serious political setback, or as an opportunity to exercise leadership in a way that builds political support for elected State and Federal Government executives. The unique political landscape at the time of the storm presented yet another challenge for the military response that would ultimately generate significant influence on the actual conduct of response operations.

RESEARCH METHODS AND STUDY OUTLINE

This monograph presents the findings of our case study. It combines personal interviews with extensive document analysis to form the substance of the report and final recommendations. We were fortunate to interview several high-ranking civilian and military officials with practical and relevant knowledge of both the evolving dual status commander conversation and the Hurricane Sandy response effort in New York. In addition to reviewing over 1,000 pages of material relevant to dual status commanders and Hurricane Sandy, we conducted 20 individual interviews and two focus groups with civilian and military personnel representing the Office of the Secretary of Defense, United States Northern Command (NORTHCOM), the National Guard, and federal armed forces. Our research represents a broad range of perspectives within State and Federal Government and offers one of the most comprehensive and detailed studies on Hurricane Sandy and the dual status commander construct to date.

Before describing the operation and presenting our findings, we offer a brief discussion of the history and evolution of the dual status commander concept. In this context, we discuss the relevant laws and constitutional authorities governing the use of military forces in a domestic environment, to include a review of the tenets of federalism and state sovereignty. We also describe the recent legislative chronology of the dual status commander initiative and its eventual adoption into federal law. After discussing the legal history and framework of the dual status commander concept, we offer a brief chronological description of the Hurricane Sandy response in New York State. We separate the response timeline into five sections (Shaping and Anticipation, Initial Response, Continued Operations, Stabilizing Operations, and Transition Operations), each describing a period of time ranging from 3-7 days. Using the description of events as a basis for further discussion, we then provide an analysis of the notable successes and areas of improvement based on our research. Finally, we conclude the monograph with a detailed discussion of 15 policy and strategy-specific recommendations intended for senior DoD leadership consideration. In summary, the monograph is structured as follows:

- Evolution of the Dual Status Commander
  - Legal Framework
  - Federalism and Sovereignty
  - Origin of the Dual Status Commander Legislation
- The Military Response to Hurricane Sandy
  - Shaping and Anticipation
  - Initial Response
  - Continued Operations
  - Stabilizing Operations
  - Transition Operations
- Post-Event Lessons Learned
  - Successes
  - Shortfalls
- Recommendations for Improvement
  - Operational and Strategic Recommendations
  - Policy Recommendations
- Conclusion

ENDNOTES - CHAPTER 1

1. Department of Defense, *DoD Directive 3025.18: Defense Support of Civil Authorities,* Washington, DC: U.S. Government Printing Office, September 21, 2012, p. 4.

2. Eric S. Blake, Todd B. Kimberlain, Robert J. Berg, John P. Cangialosi, and John L. Beven II, *Tropical Cyclone Report Hurricane Sandy (AL182012),* Miami, FL: National Oceanic and Atmospheric Administration, National Hurricane Center, 2013, p. 1, available from *www.nhc.noaa.gov/data/tcr/AL182012_Sandy.pdf.*

3. DoD generated more than 10 after-action reports specifically addressing the response to Hurricane Sandy. The Office of the Secretary of Defense, United States Northern Command (NORTHCOM), the U.S. Marine Corps, Naval Warfare Development Command, Joint Task Force Sandy, Joint Task Force-Civil Support, and others developed and published individual reports made available to the author for assistance with this research.

4. "No-notice/limited-notice" is the DoD accepted vernacular referring to incidents other than planned events of national significance (i.e., national security special events such as the Super Bowl, political conventions, etc.). According to DoD personnel, the no-notice/limited notice designation can often apply to Hurricanes, earthquakes, tornadoes, terrorism, etc. Currently, there is no doctrinal distinction between no-notice and limited-notice.

CHAPTER 2

EVOLUTION OF THE DUAL STATUS COMMANDER

While military assets can provide a valuable service during civil support operations, establishing a unified effort between Title 10 and Title 32 forces has proven to be problematic in past civil support efforts. The dual status commander (DSC) concept offers a command arrangement legally authorizing one military officer to assume simultaneous but mutually exclusive command authority over both National Guard forces under State Active Duty (SAD) or Title 32 status and Title 10 federal military forces. According to draft DoD Instruction 3025.xx, "Dual-Status Commanders for Defense Support of Civil Authorities," a dual status commander is:

> A military commander who may, in accordance with the law, serve in two statuses, Federal and State, simultaneously while performing the duties of those statuses separately and distinctly.[1]

The Government Accountability Office (GAO) defines dual status commanders as:

> Military officers who serve as an intermediate link between the separate chains of command for state and federal forces—have authority over both National Guard forces under state control and active duty forces under federal control during a civil support incident or special event.[2]

More simply stated: a DSC is "responsible for performing two separate and distinct but related jobs with two separate and distinct teams for two separate and distinct bosses, all at the same time."[3] The need for such a unique command architecture stems from the various constitutional and legal considerations governing the use of military forces in a domestic capacity, the root of which is our federalist system of government. The process used for designating a DSC is shown in Appendix I.

LEGAL FRAMEWORK

Federalism implies a system of shared powers between individual states and the national government. These powers, as related to command and control of military forces, are defined in the U.S. Constitution and establish the legal authorities and limitations for the employment of the military in domestic operations. Figure 2-1 shows the relevant sections and their relationship to each other. The figure also serves as a reference for the subsequent discussion.

In addition to authorizing Congress to "raise and support Armies,"[4] the Constitution also states:

> Congress shall have the power. . . . To provide for calling forth the militia to execute Laws of the Union, suppress insurrections and repel invasions; To provide for organizing, arming, and disciplining, the militia, and for governing such part of them as may be employed in the service of the United States, reserving to the states respectively, the appointment of the officers, and the authority of training the militia according to the discipline prescribed by Congress.[5]

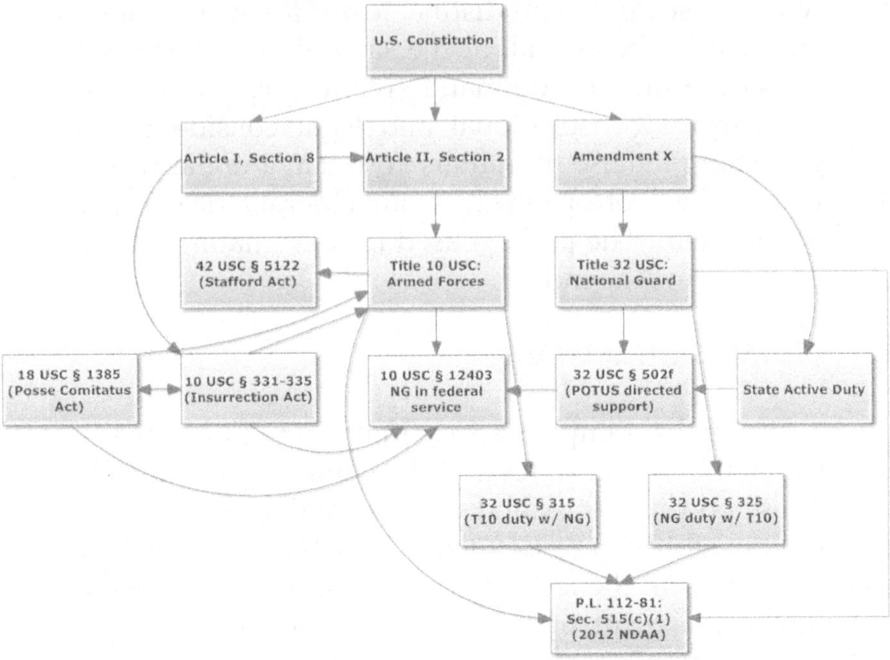

Figure 2-1. Domestic Military Law Authorities and their Relationships.

While these authorizations ensure states' rights to maintain a militia, or what is now the National Guard, the language also ensures individual states' rights are subordinate to the power of Congress under certain conditions. Article II, Section 2 of the Constitution upholds this authority by stating:

> The President shall be commander in chief of the Army and Navy of the United States, and of the militia of the several states, when called into the actual service of the United States.[6]

The federalist construct and its associated law, as applied to domestic military force operations, is intended to provide the legal mechanism for enabling a unified military response under the order of the President during incidents of national significance requiring a combined response from the National Guard and the Armed Forces. However, as we have seen in past operations, most notably in Hurricane Katrina, the Federal Government's constitutional authorities conflict with the perceived rights and responsibilities of the individual states and territories. The issue of federal control versus state sovereignty presents a significant point of friction between the States and the Federal Government that continues to challenge the effective command and control of the military, most notably with regard to the National Guard and the various duty statuses under which it serves during domestic operations. Moreover, the actual extent of emergency powers and the range of discretionary authority state governors can exercise under the 10th Amendment of the Constitution is not a well-settled area of law or public policy that needs further examination.[7]

National Guard forces, unlike the Active and Reserve components of the federal armed forces, can serve in three different duty statuses during a domestic operation. When activated in SAD status, the National Guard serves under the command of the state governor through the Adjutant General (TAG), receives state pay and benefits, and is not subjected to the restrictions of Posse Comitatus;[8] that is, they can engage in law enforcement activities when directed. When supporting operations undertaken at the request of the President or Secretary of Defense

(SECDEF), the National Guard serves under the authority of 32 United States Code (U.S.C.) § 502f; or in Title 32 status.[9] Unlike SAD, a Title 32 designation must be requested by the governor and approved by the President. Once approved, Title 32 status entitles National Guard forces to receive federal pay and benefits while remaining under command and control of the state governor. This is advantageous for operations spanning multiple states, as it eliminates the disparity in state pay rates and ensures state governors command integrity of their National Guard forces. Title 10 U.S.C. pertains to the laws regulating the Armed Forces. In accordance with the language of the Constitution, Title 10 provides the legal authority for the President to "call into actual service"[10] elements of the National Guard for federal duty. This ability to federalize state National Guard forces sets the legal precedent for the President to assume full authority over the militia.[11] While the National Guard can serve under Title 10 status, this authority is used almost exclusively in support of overseas operations. Table 2-1 summarizes the authorities and responsibility for different aspects of the National Guard under various duty statuses (SAD, Title 32, and Title 10).

Duty Status	State Active Duty	Title 32	Title 10
Command Authority	Governor		President
Pay and Benefits	State	Federal	
Posse Comitatus Act	N/A		Yes

Table 2-1. National Guard Duty Statuses.

In contrast, all Active and Reserve components of the Army, Navy, Air Force, and Marine Corps are considered federal military forces and serve under Title 10 authority. Title 10 forces, as they are referred to during civil support scenarios, receive federal pay and benefits, and are subjected to the restrictions of Posse Comitatus. These duty status distinctions are financially and legally necessary to distinguish the roles, responsibilities, and authorities between the states, Federal Government, and their respective military assets during domestic operations. The previous discussion also serves as the basis for the development of the dual status commander construct.

FEDERALISM AND SOVEREIGNTY

The interpretation of authority and legality concerning the command and control of military forces in the homeland continues to create tensions between states and the Federal Government. The conflict between state power and federal authority introduces confusion during response operations involving both the Armed Forces and National Guard. Without clearly established chains of command, lines of authority, and mission tasks, achieving unity of effort has proven difficult in past operations of large magnitude.[12] *U.S. Army Field Manual (FM) 3-28, Civil Support Operations,* articulates the command complexities between the Armed Forces and the National Guard conducting simultaneous domestic operations:

> There is not a chain of command in the military sense between the President and the Governors. The President as head of the federal government and military commander in chief may only exercise the authorities granted in the Constitution and U.S. law. Within their respective states, the Governors retain executive authority, to include command over their state's national

guard (Air and Army), until such time as the President mobilizes it for federal service. This is unique to this operational environment, and commanders at all levels need to understand the impact it has on the conduct of operations.[13]

As a result of the several duty statuses the National Guard can occupy during domestic response missions, combined with the possibility of integration with federal military assets, there are currently four command and control models for consideration when coordinating a combined state and federal response effort (see Table 2-2). As the table shows, the four models represent unique command arrangements, each of which offers advantages and disadvantages depending on one's perspective.

Command Option	National Guard	Armed Forces
State*	Governor	
Parallel	Governor	President
Dual Status	Dual Status Commander (32 U.S.C. § 315/325)	
Federal	President	

* = Conceptual model. While such a model has been proposed in past legislation, currently, there is no legal basis for the governor of a state to assume direct command authority over federal military forces.

Table 2-2. Domestic Military Command Options.

A unified and effective response is a desired end state of civil support operations involving military assistance. However, the constitutional impediments discussed earlier fuel the noted tension between states and the Federal Government during domestic response missions. Therefore, the previously discussed dual status model, wherein a single military officer commands both state and federal forces simultaneously, was created in an attempt to address some of the noted concerns.

ORIGIN OF THE DUAL STATUS COMMANDER LEGISLATION

Through the 2004 National Defense Authorization Act (NDAA),[14] Congress passed legislation allowing a National Guard officer to hold temporarily both a state and federal commission simultaneously, or serve in a "dual status."[15] Since the 2004 enactment, DSCs have been authorized to command 23 events of national significance, most notably events designated as National Security Special Events.[16] DSCs commanded operations in support of national political conventions, summits, and sporting events integrating Title 10 and Title 32 forces under a single commander utilizing separate chains of command. These were preplanned events with extensive preparation prior to execution. As such, many viewed these DSC-led operations as a success. In contrast, no-notice/limited-notice incidents like hurricanes, earthquakes, and wildfires are less certain and can lead to significant coordination challenges, such as those observed during Hurricane Katrina.

The U.S. military's response to Hurricane Katrina was widely criticized for coordination failures between state and federal levels, procedural inefficiencies, force allocation redundancy in some places and gaps in others, administrative and legal failures, and overall response timeliness.[17] In the years following the divided military response to Katrina, state governors and DoD officials realized the urgent need for policy changes and the requirement for an improved coordination mechanism between State Government, Federal Government, National Guard forces, and federal military forces. Realizing the legal precedent in place under 32 U.S.C. § 315 and 325 (a)(2),[18] a new series of conversations developed in an effort to simplify the orders process, reduce force redundancy, and close the operational gaps within the DSCA environment, all while simultaneously addressing the noted tension between state sovereignty and federalism, the legality and constitutionality of using military forces for domestic response, and the financial barriers present when combining the National Guard and Armed Forces. (See Figure 2-2.) This began a lengthy debate over legislation outlining how military forces would operate in future domestic operations.

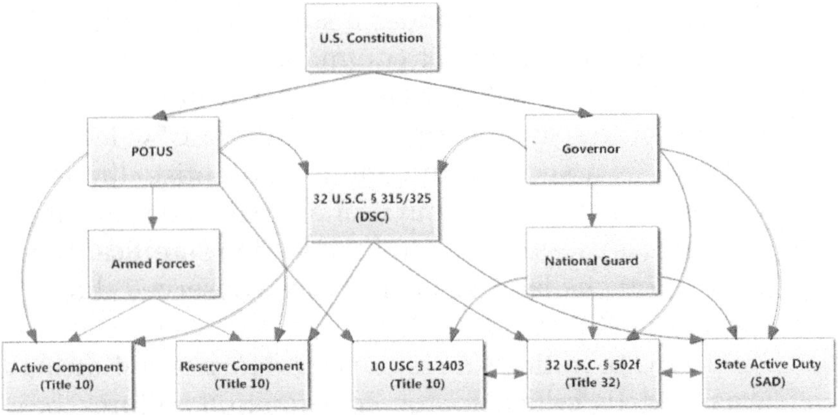

Figure 2-2. Dual Status Commander Influence Diagram.

Between 2006 and 2010, annual NDAAs contained legislation that outlined changes to the authority and control of Title 10 and Title 32 forces operating in support of no-notice/limited-notice incidents. Following repeated failed attempts between state and federal leadership to legislate a mutually agreeable command and control mechanism for emergency and disaster response, the DoD and a previously appointed Council of Governors came to an agreement on the use of dual status commanders during such events.[19] In 2010, the Joint Action Plan for Developing Unity of Effort was signed and agreed upon by the DoD and the Council of Governors representing state interests.

The Joint Action Plan effectively established the guidance authorizing a DSC, in a simultaneous but mutually exclusive manner, to command both National Guard and federal armed forces during incident response scenarios. The goal of this agreement was to establish a common operating picture between State and Federal Governments regarding the employment of military forces in response to domestic emergencies or disaster. As noted by the plan, the DSC arrangement offers alternative command architecture to the traditionally divided parallel model and provides, conceptually at least, a mechanism in which state sovereignty and federal interests can be equally balanced.[20] With the Joint Action Plan signed and agreed upon in early-2011, DSCs were authorized to command the DSCA response for both Hurricanes Irene and Isaac in 2011. However, DSCs were either not activated for these events or did not receive both Title 10 and Title 32 forces for the operations.[21] Nonetheless, these events provided a valuable opportunity for governors and DoD officials to test the DSC activation process and further endorse its concept. With the notable success of past negotiations and the noted potential of the command concept following Irene and Isaac, legislators included the DSC concept in the 2012 NDAA, which codified the DSC construct into law as the default command arrangement during incident response scenarios and specified the DSC as the "usual and customary command and control arrangement, including for missions involving a major disaster or emergency."[22] Less than 10 months after the 2012 NDAA was signed into law, Hurricane Sandy became the first time a dual status commander received Title 10 and Title 32 forces to execute unplanned civil support operations. Therefore, it is imperative that we take this opportunity to learn from the event and improve our future domestic response capabilities under the DSC arrangement. The following sections discuss the events of Hurricane Sandy in New York, with emphasis on the challenges faced under the DSC arrangement and some of the notable issues associated with the historic response.

ENDNOTES - CHAPTER 2

1. Department of Defense, DoD Instruction 3025.xx: Dual Status Commanders for Defense Support of Civil Authorities, Washington, DC: U.S. Government Printing Office, n.d (draft).

2. "Homeland Defense: DOD needs to address gaps in homeland defense and civil support guidance," Report to Senate Committee on Homeland Security and Governmental Affairs, #13-128, Washington, DC: U.S. Government Accountability Office, October 2012, p. 2.

3. Office of the Secretary of Defense. "Fact sheet: Dual Status Commander Overview," Washington DC: Department of Defense, n.d.

4. U.S. Constitution, Art. I, Sec. 8, available from *https://www.law.cornell.edu/constitution/articlei.*

5. *Ibid.*

6. U.S. Constitution, Art. II, Sec. 2, Ithaca, NY: Cornell University, available from *www.law.cornell.edu/constitution/articleii#section2.*

7. The 10th Amendment to the Constitution states: "The powers not delegated to the United States by the Constitution, nor prohibited by it to the states, are reserved to the states respectively, or to the people." This amendment ensures states maintain the rights and individual authority to govern themselves in those situations not covered in the Constitution or other federally applicable laws and regulations.

8. 18 U.S.C. § 1385: "Use of Army and Air Force as Posse Comitatus," Washington, DC. Except under extraordinary circumstances (invocation of the Insurrection Act), Posse Comitatus restricts the President from using federal military forces in a law enforcement capacity.

9. 32 U.S.C. § 502, "Required Drills and Field Exercises," Subsection (f), Washington, DC, 1964, 2006.

10. U.S. Constitution, Art. II, Sec. 2, Ithaca, NY: Cornell University, available from *www.law.cornell.edu/constitution/articleii#section2.*

11. The Militia Act of 1903 (also known as the Dick Act) codified the Federal Government's ability to federalize the National Guard under specific circumstances. The Act also provided federal funding authority for the National Guard and established the militia as two distinctive groups (in addition to the Regular Army): 1) the Organized Militia, consisting of members of the National Guard and Naval Militia; and 2) the Unorganized Militia, whose members consist of all able-bodied males over 18 years of age who are not members of the organized militia or armed forces.

12. Peter Topp, "What Should Be the Relationship between the National Guard and United States Northern Command in

Civil Support Operations Following Catastrophic Events?" Monterey, CA: Naval Postgraduate School, September 2006, pp. 1-73; Garrett Jensen, "DoD Disaster Response: Unity of Effort Versus Unity of Command," Carlisle, PA: U.S. Army War College, May 2007, pp. 1-22; Michael Teague, "The Domestic Coalition: The Command and Control Relationship between Active Component and National Guard Forces in Defense Support of Civil Authorities Operations," Newport, RI: Naval War College, June 2007, pp. 1-17; Bert Tussing, "The Potential Need to Establish an Appropriate Mechanism for the Military to Accompany and Support Civilian Components Focused on Regional Response to Catastrophe," Center for Strategic Leadership Issue Paper, Vol. 5-07, Carlisle, PA, U.S. Army War College, July 2007, pp. 1-4; Jeffrey Burkett, "Command and Control of Military Forces in the Homeland," *Joint Force Quarterly*, Issue 51, October 2008, p. 130; Andy Hall, "Improving Military Response to Catastrophic Events within the United States," Carlisle, PA: U.S. Army War College, March 2009, pp. 1-2.; William Prendergast, "Contingency Dual Status Commander: Balancing Title 10 and 32 Responsibilities," Carlisle, PA: U.S. Army War College, March 2011, pp. 1-24; Caroline Prosch, "Getting to One from Title 10 + Title 32 Unity of Effort in the Homeland," Monterey, CA: Naval Postgraduate School, September 2011, pp. 1-133; Bert Tussing, "Implementing a New Vision: Unity of Effort in Preparing for and Responding to Catastrophic Events," Center for Strategic Leadership Issue Paper, Vol. 2-11, Carlisle, PA: U.S. Army War College, March 2011, pp. 1-6; Alfred Poirier, "FEMA Urban Search and Rescue Teams: Considering an Improved Strategy for an Evolving Homeland Security Enterprise," Monterey, CA: Naval Postgraduate School, September 2012, pp. 1-105; Steven Blum and Kerry McIntyre, *Enabling Unity of Effort in Homeland Response Operations*, Carlisle, PA: Strategic Studies Institute, U.S. Army War College, April 2012.

13. Headquarters, Department of the Army, *Field Manual 3-28, Civil Support Operations*, Washington, DC: U.S. Government Printing Office, August 2010, p. 1-4.

14. Which cites similar law from 2000: 32 U.S.C. § 315, Detail of regular members of the Army and Air Force to duty with the National Guard.

15. 32 U.S.C. § 325 (a) 2), 2004, Relief from National Guard duty when ordered to Active Duty.

16. Office of the Secretary of Defense, History of dual status commander authorizations, Washington, DC: DoD, 2013.

17. Steve Bowman, Lawrence Kapp, and Amy Belasco, "Hurricane Katrina: DoD Disaster Response," Congressional Research Service (CRS) Report for Congress, Washington, DC: CRS, September 19, 2005; John Milliman, John Grosskopf, and Ozzie Paez, "Emergency Managers' Views on Improving Defense Support/ Military Assistance to Civil Authorities (DSCA/MACA) Education Programs," *Journal of Homeland Security and Emergency Management*, Vol. 3, Issue 2, 2006, pp. 1-13; Kathleen Tierney, Christine Bevc, and Erica Kuligowski, "Metaphors Matter: Disaster Myths, Media Frames, and Their Consequences in Hurricane Katrina," *The Annals of the American Academy of Political and Social Science*, Vol. 604, March 2006, pp. 57-81; Kathleen Gereski, "The Department of Defense as Lead Federal Agency," *Homeland Security Affairs*, Vol. 2, No. 3, October 2006, pp. 1-21; Topp; Jensen; Jason Schwabel, "Command, Control, Coordination, and Cooperation during Defense Support to Civil Authority Operations," Carlisle, PA: U.S. Army War College, March 2007, pp. 1-16; Teague; Burkett; Tussing, "The Potential Need," pp. 1-4; Steven Osterholzer, "Education in Action: Educating USNORTHCOM's Critical Stakeholders at the Away Game," Monterey, CA: Naval Postgraduate School, September 2008, pp. 1-91; Prendergast; Charles Dunphy, Jr., and Christopher Radel, "Assessment of Local, State and Federal Request Processes for Defense Support to Civil Authorities in the United States," Monterey, CA: Naval Postgraduate School, December 2009, pp. 1-63; Hall; Larry Porter, "Defense Support of Civilian Authorities (DSCA): What Emergency Managers Need to Know," ProQuest Dissertations and Theses, No. 305210723, The George Washington University, May 2010, pp. 1-241; Aruna Apte and Susan Heath, "Request and Response Processes for Department of Defense Support during Domestic Disasters," *Journal of Homeland Security and Emergency Management*, Vol. 8, Issue 1, 2011, pp. 1-19; Prosch; Ludwig Schumacher, "Dual Status Command for No-Notice Events: Integrating the Military Response to Domestic Disasters," *Homeland Security Affairs*, Vol. 7, No. 4, February 2011, pp. 1-9; Tussing, "Implementing a New Vision," pp. 2-4; Michael Bentley, "The Role of Military Forces in Disaster Re-

sponse: Remove the Impediments," Carlisle, PA: U.S. Army War College, March 2012, pp. 1-34; Blum and McIntyre; Paul McHale, "Critical Mismatch: The Dangerous Gap between Rhetoric and Readiness in DOD's Civil Support Missions," The Douglas and Sarah Allison Center for Foreign Policy Studies, State Resolution (SR) No. 115, Washington, DC: Heritage Foundation, August 13, 2012, pp. 1-29; "Homeland Defense DOD Needs to Address Gaps in Homeland Defense and Civil Support Guidance," Report to Senate Committee on Homeland Security and Governmental Affairs, No. 13-128, Washington, DC: U.S. Government Accountability Office, October 2012.

18. 32 U.S.C. § 315 provides the legal authority for a commissioned officer of the Active Component Army or Air Force to be detailed to temporary state duty with the National Guard. Any federal military commander assigned to state duty under this authority must receive a state commission in the appointed place of duty. National Guard officers, in contrast, hold both State and Federal (reserve) commissions and can command in both statuses. Per 32 U.S.C. § 325, National Guard officers can also be relieved from National Guard duty when ordered to Active Duty. Partly because they must also receive state commissions, Active Component officers are rarely considered for use as dual status commanders.

19. John Gereski and Chris Brown, "Two Hats Are Better Than One: The Dual-Status Commander in Domestic Operations," *The Army Lawyer*, June 2010, pp. 72-79; Prendergast; Schumacher; email communication between the author and an employee in the Office of the Secretary of Defense, May 2013.

20. "Action Memo: Joint Action Plan for Developing Unity of Effort," Washington, DC: Office of the Secretary of Defense, October 29, 2010, p. 4-6.

21. Interviews conducted by the author with various DoD employees, including those within the Office of Homeland Defense and America's Security Affairs and United States Northern Command, January-March 2014.

22. National Defense Authorization Act for Fiscal Year 2012, Public Law 112-81, H.R. 1540, p. 98.

CHAPTER 3

THE MILITARY RESPONSE TO HURRICANE SANDY

The following sections offer a brief chronological description of the events of Hurricane Sandy under Joint Task Force (JTF) Sandy in New York from October 22 to November 15, 2012. This discussion is limited to the events surrounding the dual status commander (DSC) arrangement and any associated circumstances or considerations. To provide context, the discussion begins with a review of the unique geo-political environment in the New York metropolitan area. The events are then grouped into five similarly named categories representing a defined date range:

1. Shaping and Anticipation: October 22-29.
2. Initial Response: October 30-November 2.
3. Continued Operations: November 3-5.
4. Stabilizing Operations: November 6-9.
5. Transition Operations: November 10-15.

These categories align with existing Department of Defense (DoD) reference publications detailing defense support of civil authorities (DSCA) response phases. Also, to provide context, Figure 3-1 provides maps of the study area and snapshots of the military presence in three periods: October 30-November 2; November 3-5; and November 6-9. The placement of unit symbols within the counties (circles for National Guard and pentagons for federal military) is not representative of their actual locations during the response. If military activities occurred in a particular county during the defined date ranges, we placed a single unit graphic near the center of the county to illustrate a force presence. Force strength numbers for federal forces involved in the Sandy response were either inconsistent or unavailable, so we did not scale the unit graphics for federal forces. Consistent National Guard force strength estimates, however, were available to us via daily DoD press briefings detailing the ongoing storm response. The circles representing National Guard forces are scaled in size according to the average daily force strength estimates provided for the New York metropolitan area:

- October 30-November 2: approximately 2,300 Guardsmen;
- November 3-5: approximately 3,200 Guardsmen;
- November 6-9: approximately 4,100 Guardsmen.

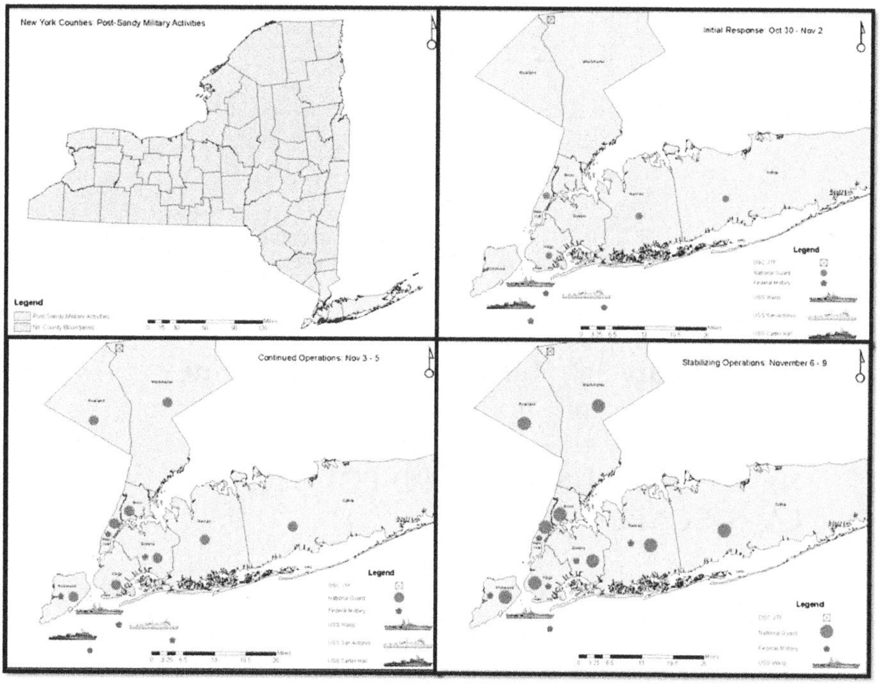

Figure 3-1. Military Activities During Hurricane Sandy.

The activities are summarized in the timeline shown in Figure 3-2.

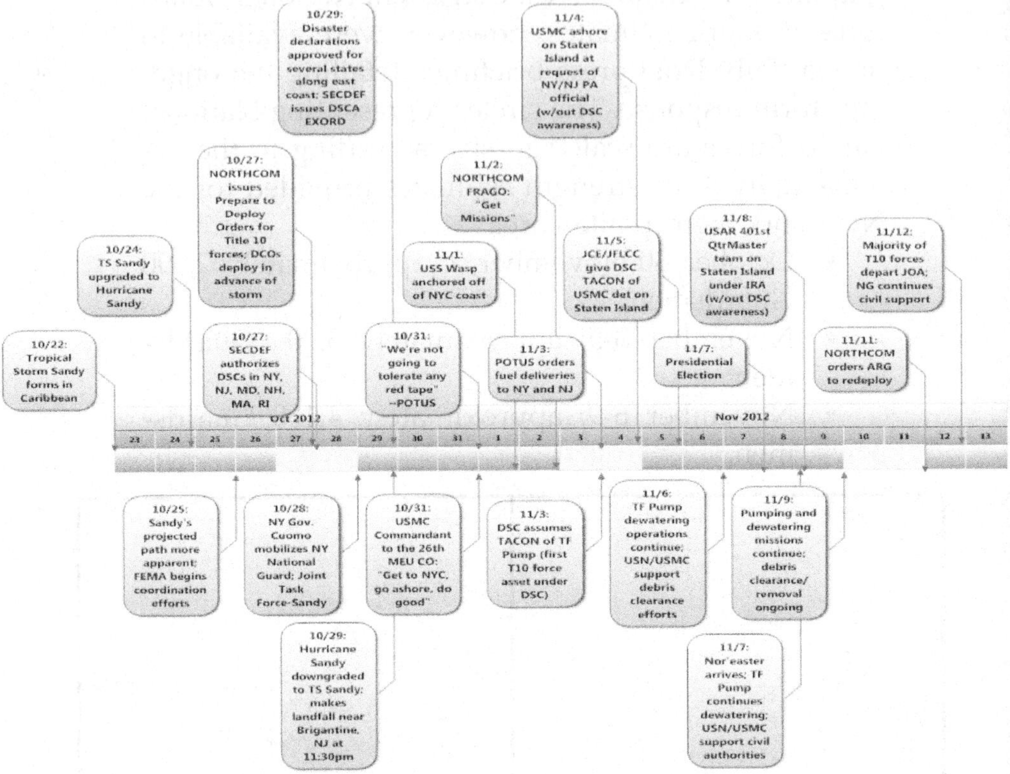

Figure 3-2. Hurricane Sandy: Timeline of Significant Events.

NEW YORK'S UNIQUE GEO-POLITICAL LANDSCAPE

Hurricane Sandy's near-direct hit on the most populated city in the United States[1] and the financial center of our economy less than 1 week prior to a presidential election was unprecedented, a coincidence noted by multiple interviewees.[2] In addition to the timing of the storm's landfall, the ensuing state and federal response was largely influenced by the region's unique geo-political structure. Given this fact, it is important to understand the organizational context of the response by discussing some of the complexities within the system of government in and around New York City.

New York is a "home rule" State. Therefore, local municipalities below the state level can, with some restrictions, create and enact laws, and govern themselves as they see fit without state legislature approval. As a city municipality, New York City is no exception to this rule. Where New York City differs is in the unique structure of its governmental leadership within its five boroughs (Manhattan, Brooklyn, Queens, Bronx, and Staten Island). Each of the five boroughs, all of which maintain separate county distinctions (New York, Kings, Queens, Bronx, and Richmond Counties, respectively), is represented by a borough president rather than a county seat. The borough presidents are elected officials and interact directly with the mayor of New York City, who serves as the representative of all five counties. In addition, the Port Authority of New York and New Jersey operates each of the airports, bridges, maritime ports, and ground transportation terminals in the New York City metropolitan area, including property in New Jersey. The complexities of the transportation network and commuting patterns within the New York metropolitan area, coupled with the diversity of its local commerce, further complicate city management functions.

Adding to the confusion is the influence of and interactions with the counties surrounding New York City's boroughs to the north (Essex, West Chester, and Rockland) and on Long Island to the east (Nassau and

Suffolk). This layered bureaucracy creates complexity in the simplest of government activities and is often influenced by state versus local politics and, in some cases, personality conflicts. Beyond this, state and federal politics do not function the way a military command structure functions. Whereas the military uses an extensive hierarchical system of command delineation, civilian leaders operate under local and state jurisdictions. Therefore, the governor cannot dictate orders to city mayors just as the President cannot dictate orders to a state governor. Coordinating a regional response effort incorporating town, borough, county, state, and federal entities across a divided geo-political landscape can prove a daunting task. As suggested by multiple interviewees, the timing of Sandy's landfall prior to the upcoming presidential election, combined with New York City's unique geo-political environment, may have prompted more aggressive involvement by the Federal Government and had a significant influence on the conduct of the joint state and federal military response activities.[3]

Shaping and Anticipation (October 22-29).

The 2012 Atlantic hurricane season was a particularly active one, with 19 named storms, 10 of which became hurricanes.[4] On October 22, 2012, the 18th tropical depression of the season formed over the southwestern Caribbean Sea and quickly strengthened into Tropical Storm Sandy late that day. On October 24, less than 2 days after its initial formation, Tropical Storm Sandy was upgraded to Hurricane Sandy near Kingston, Jamaica.[5] A day later, on October 25, Hurricane Sandy's projected path had become more apparent. As a result, the Federal Government, led by the Federal Emergency Management Agency (FEMA), began coordinating with several states in the mid-Atlantic region likely to be impacted by Sandy in the coming days. With the growing likelihood of a significant event unfolding, DoD, through the Northern Command (NORTHCOM), issued deployment preparation orders for pending DSCA operations on October 27.[6] Shortly thereafter, NORTHCOM deployed multiple defense coordinating officers (DCOs)[7] to FEMA Regions 1 (New England), 2 (Northeast), and 3 (mid-Atlantic) to assist in future DoD resource coordination efforts. While DoD coordinated its preparations, Connecticut, Massachusetts, New York, Pennsylvania, New Jersey, Delaware, and Virginia each activated National Guard troops in their states. Over a period of 6 days, states and the Federal Government went from routine operations to a heightened state of alert in preparation for the arrival of this historic storm.

As the storm approached the coast on October 28, President Barack Obama signed Stafford Act emergency declarations for Connecticut; Washington, DC; Maryland; Massachusetts; New Jersey; and New York. Over the next 24 hours, Hurricane Sandy weakened from a category 1 hurricane to a tropical storm. The storm made landfall slightly north of Atlantic City near Brigantine, NJ, at approximately 11:30 p.m. on October 29, 2012.[8] That same day, President Obama signed additional disaster declarations for Delaware, Rhode Island, and Pennsylvania; and declared major disaster areas in New Jersey and New York following massive storm surges along each coast. With the new disaster declarations approved, the Secretary of Defense, through the Chairman of the Joint Chiefs of Staff, issued a standing execution order directing NORTHCOM to provide direct support to FEMA in the affected states.[9] This would set the stage for the unprecedented combined state and federal military response under the DSC arrangement.

Owing to the Joint Action Plan and 2012 NDAA, state governors had the option to request a dual status commander for the pending DSCA response. Ultimately, six states received authorization to employ a DSC: New York, New Jersey, Maryland, New Hampshire, Massachusetts, and Rhode Island. Of the six states receiving DSC authorizations, only two—New York and New Jersey—actually activated a DSC to lead the military response efforts.[10] While the military response in New Jersey under Brigadier General Bud Grant did receive both Title 10 and Title 32 forces, this response effort was neither as geographically or politically complex as the New York response, nor did it experience the challenges noted in the DSC-led response in New York. For these reasons, the DSC-led response in New York under JTF Sandy, commanded by Brigadier General Michael Swezey, is the primary focus of this case study.

Initial Response (October 30-November 2).

As this was the first ever use of DSCs during a no-notice/limited-notice incident combining state and federal response forces, Hurricane Sandy was, undeniably, an event of national and historic significance. In addition to the significance of the military response, Sandy also led to the first 2-day closure of the New York Stock Exchange since 1888.[11] The storm also precipitated only the second mandatory evacuation ever issued for low-lying parts of New York City. In total, 375,000 people were ordered to evacuate prior to the storm.[12] Post-storm assessments suggest that over 305,000 homes were destroyed in New York, most of which were located along major coastlines and subjected to storm surge.[13] The storm surge also flooded an estimated 2,700 homes and businesses in the city[14] and rendered more than 2,000 homes on Long Island uninhabitable.[15] The storm disrupted power to more than 1.5 million New York City residents and killed 43 people statewide.[16] In total, New York experienced an estimated $19 billion dollars in damages, including $5 billion for the transportation system alone.[17] The storm had a profound effect on the New York metropolitan area that ultimately led to a large military response. The events that followed during the 2-week military response to the storm provide us with several relevant topics to consider as we attempt to improve future dual status commander-led DSCA response efforts.

In the first days following Hurricane Sandy's landfall, the DoD took a proactive, if not aggressive, forward-leaning approach in its response efforts. In anticipation of the need for military support, Governor Andrew Cuomo requested a DSC through the Office of the Assistant Secretary of Defense for Homeland Defense and America's Security Affairs (HD/ASA). Following this request and routing through the various approving authorities,[18] the decision was made to establish a dual status commander-led JTF in New York to coordinate the military response within the state. In addition, members of Joint Task Force-Civil Support, commanded by Major General Jeff Mathis, deployed to Joint Base McGuire-Dix-Lakehurst (JBMDL) in New Jersey to coordinate the multistate response effort as part of the joint coordinating element (JCE). As the commanding general of this detachment, Mathis served as the JCE to the joint force land component commander (JFLCC), Lieutenant General James Caldwell, of U.S. Army North (ARNORTH). In his capacity as the JFLCC JCE, Mathis coordinated Title 10 activities between the DSCs in New York and New Jersey and ARNORTH. As the ARNORTH commanding general, Caldwell reported directly to General Charles Jacoby, Commanding General of NORTHCOM, who subsequently reported to the Secretary of Defense, through the Joint Chiefs of Staff, and ultimately to the President. On the state side, Swezey, the appointed DSC, reported to Major General Patrick Murphy, the Adjutant General (TAG) for New York. As the New York TAG, Murphy reported directly to Governor Cuomo and represented the state's military decisionmaking authority (Figure 3-3).

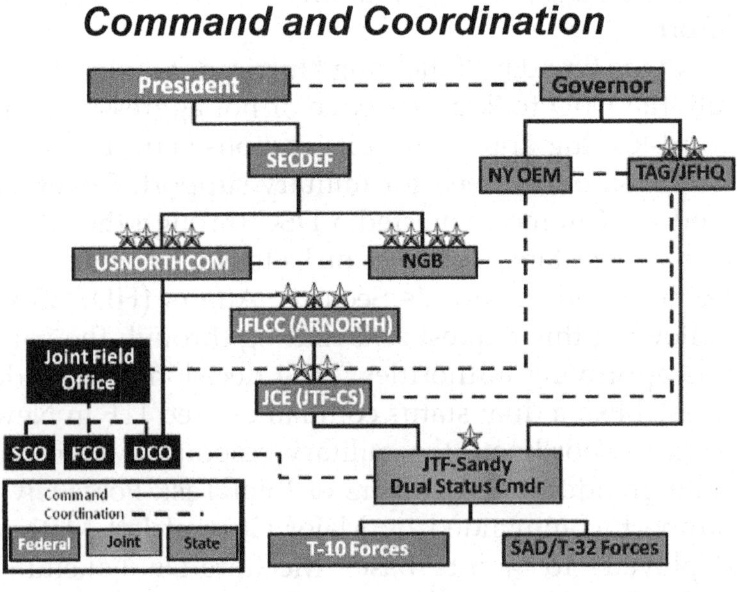

Figure 3-3. Hurricane Sandy Command Structure.

Within 2 days of Hurricane Sandy making landfall in Brigantine, NJ, the command structure had been established, and military assets from around the United States deployed to Joint Base McGuire-Dix-Lakehurst, NJ. An amphibious ready group (ARG) comprised of three U.S. Navy ships (USS *Wasp*, USS *Carter Hall*, USS *San Antonio*) sortied from Norfolk Naval Station toward the New York Harbor as part of routine hurricane avoidance maneuvers. The USS *Wasp* was the first to arrive and was anchored off the New York City coast on November 1. A reduced force contingent of the 26th Marine Expeditionary Unit (MEU) would arrive aboard the USS *Wasp* shortly thereafter. The USS *San Antonio* and USS *Carter Hall* arrived a day later on November 2.

Meanwhile, personnel from the Office of the Secretary of Defense were engaged in administrative oversight of the response while the Defense Logistics Agency, with the help of U.S. Transportation Command (TRANSCOM), began sourcing and transporting supplies to affected areas. The U.S. Army Corps of Engineers (USACE), as the lead federal agency for National Response Framework (NRF) Emergency Support Function (ESF) 3, Public Works and Engineering, was also heavily involved in the early stages of the response. Further, NORTHCOM, the National Guard Bureau, the Joint Chiefs of Staff, and other mission-critical DoD entities actively coordinated with FEMA, as well as state and local authorities throughout the mid-Atlantic region in an effort to get ahead of the storm response and provide support. This "go big, go early, go fast" approach[19] employed by DoD during the response to Hurricane Sandy ultimately would influence the remainder of the operation and provide several opportunities for lessons learned toward improving such complex operations in the future.

Additionally, the national significance of the storm on the financial center of the U.S. economy less than a week before the 2012 presidential election prompted an increased sense of urgency from the Federal Government.

> We're not going to tolerate any red tape. We're not going to tolerate any bureaucracy.
>
> President Barack Obama,
> October 31, 2012[20]

In hierarchical organizations, public statements from senior leaders and executives can greatly influence the actions of subordinates, regardless of whether such statements comply with organizational policies. In these "policy-by-speech" moments, such comments can be interpreted as standing guidance for future actions. President Barack Obama's comment after Sandy's landfall is no exception. In this case, red tape and bureaucracy can be found in national disaster guidance documents such as the NRF and National Incident Management System (NIMS). When the President publicly states that red tape and bureaucracy will not be tolerated, such guidance tends to be ignored or circumvented in order to make things happen in the most expeditious manner possible. The administration's encouragement to abandon established strategies and policies during the conduct of a domestic military response is problematic. Operational strategies guide tactical decisionmaking and are designed to accomplish a given mission in accordance with the rule of law. DoD conducts military operations in accordance with clearly established strategies and associated tactics. The military needs to know its mission, but it also needs to know the rules. Encouraging noncompliance leads to messy, chaotic, and inefficient operations.

In some cases, this is precisely what we saw during the military response in New York. The NRF, NIMS, and other such guiding documents and processes are not law. However, departing from the traditional forms of instruction tends to create additional problems as orders and missions begin to fall outside of established guidelines for conducting domestic response. In essence, these policy-by-speech moments, well-intentioned as they may be, sometimes serve the opposite intent and introduce greater confusion and/or challenges in the response.

While this is not an indictment of the President, it is also not an endorsement of the NRF and its associated guidance. National disaster response guidance is robust and detailed. Following such guidance can often be counterproductive during disasters and emergencies. However, federal disaster response is complex, so there is a reason for such formal guidance. Aggressive, mission-oriented decisionmaking by military commanders stimulated by the White House and reaffirmed by senior DoD leadership can accelerate the sometimes mechanistic response process, often leading to more effective deployment and support operations. However, this accelerated disregard for policies, procedures, and (in some cases) laws sometimes comes at the expense of unity of effort, sacrificing the principal focus for any combined state-federal response under the DSC construct.

Despite the challenges noted, military commanders and the National Guard successfully navigated the geo-political landscape of New York City. In the early stages of the military response in New York, National Guard troops conducted operations in Manhattan, and as well as Nassau, Kings, and Suffolk Counties.

New York-based National Guard units established supply points of distribution, executed search and rescue missions, assisted in resident evacuations, and supported local law enforcement by conducting security and presence patrols in areas affected by the storm.[21] In addition to coordinating the myriad initial response efforts spread across Manhattan and several surrounding counties and boroughs, personnel assigned to JTF Sandy were busy setting up an operational command center and establishing a routine for the coming days' efforts. By most accounts, the first days of the JTF Sandy response were largely effective and free of any noteworthy challenges.

As the days passed, effects of the storm compounded. Three days into the combined state and federal response, power outages still plagued areas within the five boroughs; flooding from the recent storm surge continued to hamper restoration and recovery efforts; and fuel shortages led to increasing lines at area gas stations. As a result of this, the likely influence of expanding media coverage of the storm's impacts in and around New York City, and the President's "no red tape" speech, new guidance was issued from DoD leadership to begin integrating federal military forces into the response effort. This guidance, according to sources knowledgeable on the matter,[22] was relayed from the highest levels of DoD to NORTHCOM commanders down to the tactically focused Title 10 commanders, specifically directing them to:

- Get missions;
- Do not wait for mission assignment paperwork;
- Apply total force capabilities to accomplish missions;
- When you get a mission: execute. Clean up paperwork later.[23]

This external pressure to integrate Title 10 force activity into the Sandy response, despite a lack of formal requests by New York authorities at the time, would contribute to some of the most significant activities during the entire storm response days later.

Continued Operations (November 3-5).

Nearly a week after Sandy's initial landfall, the storm's effects were becoming more apparent. Despite a range of ongoing response activities throughout the metropolitan area, there were still unmet needs noted by local officials and first responders.[24] As news media coverage grew, it contributed to the external pressures faced by the JTF Sandy staff to expand military response activities by involving prepositioned federal forces. The events of November 3-5 are among the most notable and regularly debated of the 2-week response operation.

November 3.

By November 3, National Guard forces operating in New York under Swezey were performing a range of missions in four of the five New York City boroughs, as well as four additional counties north of the city and on Long Island. While the National Guard force in New York had demonstrated its ability to meet initial requirements, there was no way to accurately predict future requirements and needed capabilities. With several Title 10 assets pre-staged at JBMDL and pressure to integrate federal forces mounting, the DSC found himself in a unique position that required balancing political influence, operational requirements, financial considerations, and legal nuances to determine the most efficient and effective manner to respond to a growing need.

On the morning of November 3, flooding from Sandy's storm surge continued to present significant challenges for storm responders. In addition, there were widespread fuel shortages due to the ongoing power outages. With 4 days remaining before the election, national news media coverage regularly broadcast footage of lengthy lines of those waiting for gasoline at area stations.[25] That morning, the White House, without consent of the states, issued an executive order for the Defense Logistics Agency to begin transport and distribution of fuel in both New York and New Jersey.[26] As a result of this new order and the increasing external pressure to involve Title 10 forces, Swezey considered deploying a contingent of Active Duty forces to assist in dewatering operations in area subways as well as increasing fuel distribution in the surrounding boroughs.[27] Prior to requesting Title 10 force support and becoming the first DSC to assume command and control of Title 10 and Title 32 forces for a no-notice/limited-notice incident, Swezey had to weigh several considerations, not the least of which were the politics influencing the response.

According to doctrine, Title 10 forces should only be considered during domestic response when local and state capacities have been overwhelmed or when civil authorities are otherwise incapable of performing the necessary mission[28]—in this case, pumping thousands of gallons of water out of subways and other flooded facilities and distributing fuel via military transport. Largely due to costs associated with using Title 10 assets, as well as the infringement on state sovereignty, federal forces, at least doctrinally, operate on a "last in, first out" philosophy in these situations. As a result, there is redundancy built into the state emergency response process. One form of redundancy regularly used during disaster response is the Emergency Management Assistance Compact (EMAC). Under EMAC, can request additional support (such as National Guard forces) from surrounding states to assist in incident response operations within their state.[29] In order for the DSC to request support from Title 10 forces, standing state-to-state EMAC agreements should be fully implemented. However, it is often left to the governor's subjective assessment to determine the point at which EMAC agreements are no longer an option, and DSCA is a requirement. Therefore, it is difficult to clearly define the appropriate trigger for requesting DSCA. This can be problematic for several reasons.

According to some, requesting Title 10 force support prior to exhausting all EMAC options carries strategic implications for the National Guard.[30] If, instead of using an EMAC to request additional National Guard troops, Title 10 forces deploy to support state operations, the resulting public perception of the National Guard may be one of ineptitude. Such perceptions can strain the relationships between the DSC (who is most often a National Guard officer) and TAG (also a National Guard officer and appointed by the governor in most states). While these are political and policy issues, they can and do influence the conduct of DSCA operations. As we saw during Sandy, the DSC had to balance the political desires of elected officials with the necessity of mission accomplishment. As a commander with two distinct chains of command, the DSC must balance both state and federal responsibilities in a way that facilitates efficient and effective tactical leadership.

With the political implications aside, EMAC requests for additional National Guard forces carry other logical considerations. Since National Guard personnel are civilians first, many have jobs outside of their role in the Guard. To justify a request for activating more Guard personnel, there must be missions to fulfill. Otherwise, activating civilian Guardsmen to wait idly by not only interrupts occupational continuity, but is also a waste of taxpayer money. During the early response to Sandy, the DSC knew he had Title 10 forces at JBMDL ready to support the response operation if requested. With the experiences of Katrina in mind, no one at the state or federal level wanted to be late to respond or be short on resources. Owing to these issues and in addition to the state of operations on the morning of November 3, Swezey, with the support of both state and federal chains of command, made the decision to deploy Title 10 forces to assist in dewatering operations in New York City. At 11:28 a.m. on November 3, 2012, Swezey became the first dual status commander in U.S. history to assume tactical control of Title 10 forces during a no-notice/limited-notice DSCA response.[31] On this day, a joint force of Army, Navy, Air Force, and Marine Corps personnel assigned to the 19th Engineer Battalion in Fort Knox, KY, deployed to the area as part of Task Force (TF) Pump and began dewatering operations under the command of the DSC and JTF Sandy.

November 4.

Less than a day removed from the successful coordination and first-time deployment of Title 10 forces under a DSC-led no-notice/limited-notice DSCA response, the dual status commander was unaware of a

Title 10 force operation ashore on Staten Island. By the evening briefing, what was a successfully coordinated operation experienced its first and perhaps most significant coordination challenge of the entire Sandy response effort.

The events of November 4 have been addressed in numerous after-action reports and post-Sandy analyses of the DSCA response. According to these sources, Swezey was made aware of the fact that U.S. Marines assigned to the 26th MEU and aboard the USS *Wasp* off the coast of Breezy Point, NY, came ashore on Staten Island in order to support local authorities. What we do not know is exactly how the Marines were requested to support civil authorities or who generated the request. While the circumstances leading to the Marine Corps' arrival on Staten Island remain in question, one thing is certain: the DSC did not request Title 10 forces to come ashore on November 4, nor was he aware of the Marines' activities until long after they arrived. The events of this day provide perhaps the single most valuable example of confusion and, consequently, opportunities for lessons learned from the entire Sandy response.

Upon learning of the Marines' landing on Staten Island, the DSC contacted his state and federal chain of command to inquire about the mission request and authorization, or lack thereof. According to sources knowledgeable on the situation, none of the general officers within the state or federal chains of command were aware of the U.S. Marine Corps (USMC) mission on Staten Island or knew who authorized the landing.[32] Fearing the perceived violation of state sovereignty as a result of a federal military force operating ashore without the governor's request or approval, the DSC requested that the Marines on Staten Island cease all operations and return to the ship. Since neither the DSC nor the governor formally requested the Marines' assistance, nor were any of the military commanders aware of the circumstances leading to the Marines' activities, it was thought at the time that this would set a bad precedent for future DSC-led DSCA response efforts. Unless justified under Immediate Response Authority per DoD Directive 3025.18, some suggested that an unauthorized Title 10 operation would question the efficacy of the newly agreed upon DSC architecture for no-notice/limited-notice response scenarios. Due to the concerns voiced by members of JTF Sandy, NORTHCOM issued guidance on the evening of November 4 to halt all Title 10 activities outside of the DSC's awareness.[33] While some voiced their concerns over the Title 10 presence on Staten Island, others praised the decision to bring Marines ashore. Regardless of position, the next challenge for the DSC was determining how and why the Marines were requested to come ashore and then, since they were ashore and capable of assisting, determining how to best use their force capability to help the citizens of New York.

November 5.

With operations continuing overnight and into the morning of November 5, the Marine Corps presence on Staten Island remained a primary focus of the DSC and other senior leaders. Other than operating under Immediate Response Authority (IRA) in accordance with DoD Directive 3025.18, there is a lengthy and often arduous request process governing how Title 10 forces receive and fill mission assignments to support civil authorities. Given the questionable circumstances of the Marines' arrival—particularly if not justified under IRA—it seems that the request for assistance (RFA) process was not followed. Despite this, the Marines were still ashore and actively involved in debris clearance and other assistance activities. After discussing options with TAG and the JFLCC, the generals determined that the DSC would assume tactical control of the Marine detachment operating on Staten Island. As the Title 10 JCE to the JFLCC, Mathis specified when the Marines were aboard the USS *Wasp*, they would be under tactical control of the joint force maritime component commander. However, in order to alleviate further confusion, the Marines would operate under the tactical control of the DSC anytime they were ashore. The JFLCC further clarified that all Title 10 forces ashore in New York must have approval and awareness of the DSC moving forward or must be performing functions under ESF 3 and in direct support of the USACE.

After clarifying these issues and assuming tactical control of the Marines in his area of operations, the DSC issued instructive guidance to the detachment authorizing them to perform debris clearance only—not debris removal, due to important Stafford Act and associated legal distinctions between the two terms.[34] Local residents welcomed the Marine Corps presence on Staten Island while local and national media outlets provided extensive coverage of the response activities. What started as a significant complication seemingly undermining the authority of both the governor and the DSC evolved into a mutually supportive and beneficial operation between the Title 10 forces and the DSC-led JTF.

Beyond the scope of the USMC activities on Staten Island, the remaining operations on November 5, consisted mainly of dewatering missions under TF Pump in the Rockaways and Manhattan. These missions continued into the evening of November 5 with visibility and approval of the DSC. At the conclusion of the day on November 5, the DSC-led DSCA response in New York under JTF Sandy now had a full contingent of both state National Guard and federal military forces operating throughout. Many of the previously encountered challenges had been resolved, or at least stabilized. The impacts of the military response effort were becoming tangible as time progressed further from initial landfall. The DSCA operation moved into the stability operations phase of the response with a positive outlook for the remaining stability and transition.

Stabilizing Operations (November 6-9).

November 6.

On the morning of November 6, the National Weather Service issued nor'easter warnings for a large area in the northeastern United States, including New York City. While both the USS *San Antonio* and USS *Carter Hall* vacated the area to avoid the coming storm, the USS *Wasp* chose to remain at anchor off the coast of the city. With a small detachment of Sailors assisting in dewatering missions on Liberty Island and Marines from the 26th MEU still supporting debris clearance on Staten Island, the ship's captain chose to weather the storm off the Breezy Point coast. With the Navy ships moving out, dewatering operations in support of TF Pump continued across the region under the tactical control of the DSC. Meanwhile, outside of the DSC chain of command, FEMA, the Defense Logistics

Agency, the USACE, and US TRANSCOM continued supporting response operations in affected areas in New York and New Jersey.[35] As the storm closed in on the New York City area, NORTHCOM officials issued guidance to all Title 10 forces supporting the response to prepare for retrograde after the storm passed. The next 2 days would see minimal changes in activity as the storm approached.

November 7.

As the nor'easter approached on November 7, Title 10 and Title 32 forces continued supporting civil authorities in a variety of capacities. According to the DoD's daily Pentagon briefing on November 7, Marine Corps engineers assigned to TF Pump assisted in dewatering operations in Queens, Air Force teams operated in support of the New York City Fire Department in the Rockaways and Brooklyn, and Navy dive teams assisted in pumping missions at the World Trade Center.[36] The DoD also reported that Marines continued to assist with debris clearance on Staten Island and pumping operations in Breezy Point alongside Navy personnel.[37]

November 8.

November 8 consisted of much of the same from the previous day's work. While the nor'easter had not completely cleared the region and continued to limit flight operations, military forces maintained their support of local authorities in dewatering missions throughout Manhattan and the surrounding boroughs. Again, according to DoD reports on November 8, elements of the Army Reserve's 401st quarter-

team supported pumping operations on Long Beach Island and Staten Island. Interviews with Sandy commanders revealed that the Army Reserve unit on Staten Island was operating outside of the DSC's awareness and under Immediate Response Authority, according to their unit commander.[38] As a result of this and the expiration of the 72-hour time limit granted under IRA, the Reserve unit was instructed to vacate the area.

While minimal in comparison to the Marines' so-called "invasion" of Staten Island, this was still a noteworthy issue that offers support for the recommendations that follow. With regard to the Marines, they continued their support to civil authorities by dewatering homes and apartment buildings in Breezy Point and the Rockaways. Airmen supported pumping missions at the Rockaway Waste Water Treatment Facility while Navy divers maintained their support of dewatering efforts at the World Trade Center.[39] Army and Air National Guard personnel continued their assistance through ongoing food and water distribution, fuel distribution, sheltering, debris removal, and donations distribution.[40] Operations continued throughout the day and into the evening without any notable incidents. By the evening of November 8, the nor'easter had passed, making way for a new day of unrestricted response efforts.

November 9.

By November 9, DSCA stability operations were nearing an end. The storm had passed, and many of the same activities from November 8 carried over into operations the following day. In addition to ongoing debris clearance and removal in Staten Island and the Rockaways, operations on November 9 included Air Force support in the Rockaways and on Long Island, Army and Navy dive team support at the World Trade Center, and Marine Corps and Navy pumping missions in the Rockaways and Breezy Point.[41]

Transition Operations (November 10-15).

Operations from November 10-15 quickly reduced in frequency and scope. By November 11, NORTHCOM had released a redeployment order for the ARG to return to Norfolk.[42] The next day, most of the 26th MEU redeployed back to Camp Lejeune, NC. The majority of the Title 10 forces departed by November 13, leaving mostly National Guard personnel in the area of operations. Seemingly as fast as the operation began, it was nearing its conclusion. By the middle of November, nearly 2 weeks after Sandy's initial landfall, most of the region's power was restored; well over a million gallons of water had been pumped from area homes, apartments, subways, and other facilities; thousands of rations of food and water were distributed, and countless quantities of debris removed from areas with damaged infrastructure. By many accounts, the first ever use of a dual status commander-led no-notice/limited-notice DSCA response was nearing a successful completion.

In total, it is impossible to say how many lives were saved due to JTF Sandy's actions during the 2 weeks following the storm. While lives saved cannot be measured, military actions during the storm response contributed in significant ways to preventing suffering and mitigating further property damage for the residents of New York and other surrounding states. The DSC-led response under JTF Sandy in New

York successfully integrated National Guard and federal armed forces for the first time in a no-notice/limited-notice incident. As with any first-time experience, there were instances of success and challenge. As a test case for future operations, this event provided several examples of lessons learned, which can be used to improve future DSC missions in similar capacities. The following section discusses some of the most pressing lessons learned, including successes and perceived failures, and analyzes the circumstances surrounding each occurrence. The lessons learned provide the foundation for the final section of recommendations based on the conclusions found in this research.

ENDNOTES - CHAPTER 3

1. See "Annual Estimates of Resident Population for Incorporated Places of 50,000 or More, Ranked by July 1, 2013 Population: April 1, 2010 to July 1, 2013-United States-Places over 50,000 Population," details available from *factfinder.census.gov/faces/tableservices/jsf/pages/productview.xhtml?src=bkmk.*

2. Interviews conducted by the author with various DoD employees, National Guard officers, and active military officers that participated in the response to Hurricane Sandy in some capacity, January-March 2014.

3. *Ibid.*

4. National Oceanic and Atmospheric Administration (NOAA), "2012 Atlantic Hurricane Season," January 23, 2014, at *www.nhc.noaa.gov/data/tcr/summary_atlc_2012.pdf.*

5. *Ibid.*, p. 4.

6. See "The Department of Defense Prepares for Hurricane Sandy," Release No: 854-12, October 27, 2012, available from *www.defense.gov/releases/release.aspx?releaseid=15646.*

7. A DCO is a single point of contact for domestic emergencies who is assigned to a joint field office to process requirements for military support, forward mission assignments through proper channels to the appropriate military organizations, and assign military liaisons, as appropriate, to activated emergency support functions. See *Joint Publication (JP) 1-02, Department of Defense Dictionary of Associated Military Terms*, p. 67.

8. National Oceanic and Atmospheric Administration.

9. For details, see Hurricane Sandy: Timeline, Washington, DC: Federal Emergency Management Agency, available from *www.fema.gov/hurricane-sandy-timeline*.

10. Interviews conducted January-March 2014; DoD Release No: 854-12.

11. Eric S. Blake, Todd B. Kimberlain, Robert J. Berg, John P. Cangialosi, and John L. Beven II, Tropical Cyclone Report Hurricane Sandy (AL182012), Miami, FL: National Oceanic and Atmospheric Administration, National Hurricane Center, February 12, 2013, p. 18, available from *www.nhc.noaa.gov/data/tcr/AL182012_Sandy.pdf*.

12. See "Mayor Bloomberg Issues Order for Mandatory Evacuation of Low-Lying Areas as Hurricane Sandy Approaches," available from *www1.nyc.gov/office-of-the-mayor/news/377-12/mayor-bloomberg-issues-order-mandatory-evacuation-low-lying-areas-hurricane-sandy*.

13. Blake *et al.*, p. 18.

14. Linda Gibbs and Caswell Holloway, *Hurricane Sandy After Action Report and Recommendations to Mayor Michael R. Bloomberg*, New York: New York City Office of the Mayor, May 2013, p. 12.

15. Blake *et al.*, p. 18.

16. Gibbs and Holloway, pp. 1, 13.

17. Blake *et al.*, p. 18.

18. See "Dual Status Commander Designation Process" flow chart in Appendix II.

19. Interviews conducted January-March 2014.

20. Quote from the President's address to residents of Atlantic City 2 days after the storm's landfall. See "Obama, Christie Tour Hurricane Ravaged NJ," available from *www.msnbc.com/the-last-word/obama-gov*.

21. Interviews conducted by the author with various National Guard officers who participated in the Joint Task Force Sandy response effort, February 2014.

22. Interviews conducted by the author with various DoD employees, National Guard officers, and active military officers that participated in the response to Hurricane Sandy in some capacity, January-March 2014.

23. Copied material from NORTHCOM guidance (fragmentary order) of November 2.

24. Interview conducted by the author with a National Guard officer that participated in the response to Hurricane Sandy, February 2014.

25. See "A Slow Return to Normal Skips the Gas Station," *The New York Times*, November 4, 2012, available from *www.nytimes.com/2012/11/04/nyregion/gas-rationing-is-new-burden-after-hurricane-sandy.html?_r=0.*

26. Interview conducted with a National Guard officer that participated in the response to Hurricane Sandy, February 2014.

27. Interviews conducted with various DoD employees, National Guard officers, and active military officers that participated in the response to Hurricane Sandy in some capacity, January-March 2014.

28. Department of Defense, *Joint Publication 3-28: Defense Support of Civil Authorities*, Washington, DC: U.S. Government Printing Office, July 31, 2013, pp. viii-ix.

29. See "What is EMAC?" available from *www.emacweb.org/index.php?limitstart=0.*

30. Interviews conducted with various DoD employees, National Guard officers, and active military officers that participated in the response to Hurricane Sandy in some capacity, January-March 2014.

31. Interview conducted with a National Guard officer that participated in the response to Hurricane Sandy, February 2014.

32. Interviews conducted with various DoD employees, National Guard officers, and active military officers that participated in the response to Hurricane Sandy in some capacity, January-March 2014.

33. "Hurricane Sandy Disaster Response Active Collection Lessons Learned," Norfolk, VA: U.S. Fleet Forces Command (USFFC), February 22, 2013.

34. The Robert T. Stafford Disaster Relief and Emergency Assistance Act (Stafford Act), Secs. 403(a)(3)(A), 407, and 502(a)(5), specify conditions when the President may authorize federal agencies to conduct debris removal or reimburse private contractors for such services using a 75/25 federal/state cost share, respectively. Under 42 U.S.C. § 5107b(a)(3)(A), federal agencies may, at the direction of the President, provide assistance to save lives and protect property, including reimbursement for debris removal services. 42 U.S.C. § 5173, supported by 42 U.S.C. § 5192(a)(5), specifies that federal agencies can conduct debris **clearance** ((a)(1)) and authorizes private contractors to conduct debris **removal** ((a)(2)). Due to these provisions, the DSC chose to authorize Marines to engage in debris clearance activities only. Also of note: Prior to October 27, 2012, 44 CFR 206.228(a)(2), "Force Account Labor," did not permit reimbursement of straight or regular time salary costs for contractors performing debris removal under the aforementioned provisions; they were instead reimbursed at the overtime rate. Following Hurricane Sandy, a retroactive revision to this law was signed allowing for reimbursement of debris removal activities only at the regular salaried rate. Debris clearance activities were not included in this revision. See "FEMA Recovery Fact Sheet 9580.215: Hurricane Sandy: Debris Removal Force Ac-

count Labor Costs," available from *www.fema.gov/media-library-data/20130726-1857-25045-0130/rp9580_215_debris_removal_force_account_labor_costs_fact_sheet__11_5_12_.pdf.*

35. See "Pentagon Officials Detail Continuing Storm Response Support," Daily Briefing, November 6, 2012, available from *www.defense.gov/news/newsarticle.aspx?id=118464.*

36. See "Pentagon Officials Provide Storm Response Update," Daily Briefing, November 7, 2012, available from *www.defense.gov/news/newsarticle.aspx?id=118473.*

37. *Ibid.*

38. Interviews conducted with National Guard officers that participated in the response to Hurricane Sandy, February 2014.

39. See "Pentagon Provides Storm Response Update," Daily Briefing, November 8, 2012, available from *www.defense.gov/news/newsarticle.aspx?id=118488.*

40. *Ibid.*

41. See "Pentagon Provides Sandy Response Update," Daily Briefing, November 9, 2012, available from *www.defense.gov/news/newsarticle.aspx?id=118496.*

42. USFFC, p. 25.

CHAPTER 4

POST-EVENT LESSONS LEARNED

Hurricane Sandy caused a great deal of damage in the New York metropolitan area. However, Sandy was only a Category 1 storm when it made landfall and quickly dissipated after coming ashore. While the storm surge was one of the most significant in New York's history, the storm could have been worse. The post-event lessons learned from this storm cover everything from conflicting command intent, command and control, communication, coordination, mission assignments, laws, policies, and even politics. Within the observations of this analysis, there are examples of successes that should be repeated, as well as examples of issues needing improvement. This chapter is divided into two main categories: successes and shortfalls. Within successes, there are four sub-categories and associated discussions:

- Coordination;
- Liaison officers;
- Forward-leaning strategy;
- Sustaining successes.

Within shortfalls are six sub-categories and associated discussions:

- Process Integrity;
- Title 10 Awareness of the Dual Status Commander Construct;
- Command Stucture;
- Mission Assignment Process;
- Defense Support of Civil Authorities (DSCA) Education;
- Dual Status Commander Guidance/ Instructions.

SUCCESSES

Between extensive dewatering and supply transport and delivery, the U.S. Army Corps of Engineers (USACE) and the Defense Logistics Agency (DLA) were critical to the ongoing success of the overall federal response. Despite the importance of these contributions to the operational successes, the dual status commander (DSC) was not in a position to command or direct USACE or DLA activities. Since these Department of Defense (DoD) activities occurred outside of the DSC-led response in New York, they are not detailed in this analysis. This is not to say that the dual status commander (DSC)-led joint task force (JTF) did not succeed. Despite being the historic first-time use of a DSC for a combined state and federal response, there were some notable successes. Hurricane Sandy provided a proof-of-concept environment for evaluating a DSC operation that involved effective coordination, the successful integration of liaison officers (LNOs), and a strategic forward-leaning approach to the operation, including pre-positioning Title 10 assets, all of which should be repeated and leveraged again in future DSCA response operations of similar circumstances and requirements.

Coordination.

While it cannot be empirically proven through systematic analysis, according to several accounts of the response effort, the DSC JTF successfully coordinated many complex staff integration processes that facilitated effective joint communication and coordination between Title 10 and Title 32 staff representatives.[1] During the beginning stages of establishing the temporary JTF, National Guard and U.S. Northern Command (NORTHCOM) staff officers transitioned into the initial operations phase with minimal complication. Initial staffing procedures, including the identification and pre-deployment of defense coordinating officers (DCOs) to the anticipated affected areas, were executed with clarity and focus. The JTF established the required staff cells and began coordinating the response operation. These critical staff procedures, at least in the early stages of the storm response, were efficient and effective, and should be used as a guide for future DSC-led JTF augmentation. Additionally, the willingness of commanders and senior leaders to use verbal orders of the commanding officer (VOCO) as a basis for executing tasks and missions was also an effective coordination mechanism noted during the Sandy response.

With the noted complexities and burdensome nature of the mission assignment process, leaders encouraged their subordinates to obtain VOCO as a basic form of approval prior to conducting response activities. Commanders and other senior leaders demonstrated a willingness and ability to coordinate verbally and direct the tactical response activities without waiting for the often sluggish written approval process to occur. In many cases, this led to quicker response activities that ultimately benefitted the citizens of New York. One specific example of this VOCO process occurred immediately following the Marine Corps' arrival on Staten Island on November 4.

Aside from the administrative coordination successes noted previously, one of the most significant tactical coordination successes that can offer insight into future command decisionmaking occurred when the DSC received word of the Marines' unsolicited (from the DSC JTF) landing on Staten Island. As mentioned earlier, following an initial period of frustration over the landing and subsequent verbal coordination with general officers within the chain of command, the DSC was given tactical control of the Marine detachment ashore instead of having them return to the ships. This decision provided two benefits: 1) the command authority of the DSC and ultimately the Governor of New York remained intact by assuming tactical control of the unrequested force; and 2) it offered a tried and accepted decision mechanism for future operations where DSCs can request tactical control of all Title 10 forces entering the joint operations area (JOA) regardless of pretext.

Conversely, had the Marines been permitted to continue operating ashore outside of the DSC command architecture, the sovereignty of New York and the governor's authority, and thus the purpose of establishing a DSC as a principal coordination mechanism between the states and Federal Government, would have been undermined. In contrast, had the Marines returned to the ship as was originally proposed, this would simply serve to further the divide between the State and Federal Government, ultimately to the detriment of the citizens of New York, further questioning the efficacy of the DSC arrangement.

As it occurred, the DSC's assumption of tactical control over the Marines, to the satisfaction of both the Title 10 and Title 32 commanders, resulted in a successful solution to what was one of the biggest points of friction and areas for improvement of the entire DSCA response to Hurricane Sandy. Assessing the dynamics of VOCO, including the level of compliance and implementation, is nearly impossible from a lessons learned perspective. However, the benefit of VOCO is evident from the Sandy response. Coordinating operational continuity by assuming tactical control of the Marines occurred almost entirely through verbal discussion and coordination, thus reaffirming the value of the VOCO process as a critical coordination practice. In many cases, VOCO coordination is facilitated by and through strategically positioned LNOs inside the relevant agencies, services, departments, and organizations.

Liaison Officers.

Interservice/agency coordination is critical to the success of any joint operation. It is perhaps even more critical to the success of no-notice/limited-notice response, given the dynamic and evolving nature of such operations. Since information and requirements so often change during these events, generating and maintaining situational awareness is a necessity for commanders. As designated representatives of their respective service or organization, LNOs and/or emergency preparedness liaison officers (EPLOs) provide a vital function to any commander executing joint operations. The placement of LNOs across the entire JOA was considered by many senior commanders to be one of the most beneficial practices employed by the joint force during Hurricane Sandy.[2] By embedding LNOs representing various services in and around important staff elements, such as the joint field office (JFO), JTF Headquarters, the Federal Emergency Management Agency (FEMA), and the respective military service headquarters, operational decisionmaking processes were enhanced through shared situational awareness.

While there were some LNO coordination gaps noted (discussed later in this monograph), the emphasis on using LNOs to improve coordination is a success worth repeating. While it is unclear exactly how many LNOs were essential to amplify command coordination during Sandy, it appears from the available data that LNO integration provided an intangible but genuine benefit. By identifying needs and assigning LNOs to critical areas, the joint force demonstrated a forward-leaning, assertive approach to this civil support operation that is also worth noting.

Forward-Leaning Strategy.

Exercising command initiative by deploying LNOs throughout the JOA is an example of a successful, forward-leaning approach employed by DoD and the National Guard in response to Hurricane Sandy. Rather than waiting to deploy forces until after receipt of a support request, both DoD and the National Guard took a proactive approach and prepositioned forces and equipment in and around areas affected by Sandy. As discussed, the lessons of Hurricane Katrina remain embedded in the minds of many. Given the highly criticized federal response to Katrina, the prevailing philosophy of senior leaders involved with the planning and execution of the Sandy response held that DoD should take aggressive measures to ensure that the citizens of New York and other affected states receive the assistance they need when they need it. In other words: "don't be late."[3]

DoD's effort to avoid repeating the failures of Katrina resulted in the aforementioned "go big, go early, go fast" approach that involved prepositioning equipment and forces and taking additional steps to facilitate coordination and communication between all participating units and agencies. Rather than operating via a pull philosophy, in which DoD waits for requests from local and state agencies to mobilize resources, a push mentality was employed. This push vs. pull approach is a paradigm shift of sorts for DoD compared to past response efforts. Placing personnel and equipment assets on standby status in geographic proximity to the JOA offered the DSC additional capabilities to consider during the response, which ultimately proved beneficial.

Although the prepositioned forces assume an associated cost risk if they are not used—and therefore pressure commanders to employ mobilized assets—the benefit of having prepositioned Title 10 forces near the JOA as a force multiplying capability is significant. Additionally, this option saves National Guard forces from activating troops, forcing them to unnecessarily leave their civilian jobs for extended periods in the event they are not involved in the response operation. In the case of Hurricane Sandy, prepositioned Title 10 forces gave the DSC the flexibility to maximize the National Guard assets currently in the JOA and prevented the unnecessary mobilization of potentially thousands of additional Guard troops at the additional expense to the taxpayer. The Title 10 forces prepositioned in and around the JOA provided an obvious benefit to the DSC that should be considered an administrative and strategic best practice for future civil support scenarios.

Sustaining Successes.

While not a comprehensive list, the aforementioned information reveals some notable successes that should be considered by those directing future operations. Sustaining the successes will provide some of the critical components of the strategic, operational, and tactical level frameworks necessary to ensure a successful DSCA response under a DSC-led JTF. To ensure the continued use and implementation of those actions for future operations, it is important to include these successes in lessons learned and after-action reports. DoD has a robust lessons learned program within each service component. Often, reports generated by Centers for Lessons Learned provide DoD with valuable information and recommendations to incorporate into future training exercises, simulations, and actual operations. Integrating these successes into the ongoing lessons learned process would ensure leaders have the information necessary for improved decisionmaking during future DSCA events.

Leveraging lessons learned ensures commanders can incorporate valuable knowledge into critical training and exercise simulations. These simulations often provide military forces with the most comprehensive and realistic opportunities to train and prepare for likely operational situations. Integrating this knowledge into future DSCA training events offers commanders an opportunity to test and evaluate the efficacy of the aforementioned strategies and tactics. By simulating such operations, commanders are better positioned to execute actual operations when the time comes. Further, training and simulations create evaluation scenarios that help identify mission-critical gaps and areas for improvement, such as training more LNOs to serve in this necessary function.

Finally, DoD and the National Guard can ensure sustained successes in future DSCA operations by training more personnel to serve as LNOs. The LNO requirement is essential for a successfully coordinated response, especially one involving multiple services, agencies, departments, and organizations. Maintaining a cadre of trained personnel capable of serving as LNOs is necessary for continued success. With trained LNOs involved in extensive exercises and simulations designed in consideration of past lessons learned, DoD and the National Guard can sustain the notable successes from Hurricane Sandy and improve performance during the next no-notice/limited-notice incident requiring DSCA. However, beyond sustaining the successes gleaned from Hurricane Sandy, there are several areas for improvement to learn from as well.

SHORTFALLS

Despite some important successes, it is crucial to note that temporary JTFs for no-notice/limited-notice incidents are just that: temporary. These makeshift commands stand up in response to events requiring joint coordination of military activities in support of civil authorities. They do not train for months in preparation for deployments, as defined military units often do. Due to this temporary joint structure, the JTF often lacks continuity and sound working relationships. As a result, these operations inevitably experience challenges. While the preceding successes offer valuable insight into sustaining future actions, there are, as expected, numerous areas for improvement worth noting.

The coordinated federal response to Hurricane Sandy had many successes; and, as is to be expected with the first-time implementation of a new command arrangement, the operation had many failures from which to learn. The following section identifies some of the more significant challenges the DSC-led JTF and associated personnel experienced during the

2-week response. For clarity and consistency, the topics are again separated into categories with a detailed description of the circumstances for context and consideration. This section serves as the basis for the final section of this monograph that details specific recommendations for improvement of strategy and policy.

Process Integrity.

As previously discussed, the preparation phase of the Sandy response was efficient and largely effective. In New York, civilian and military personnel deployed to the planned JOA early and set up a functioning JTF ready to coordinate the receipt and employment of state and federal forces. From the storm's landfall on October 29 to the conclusion of initial area assessments on October 31, most accounts of the operation were positive. National Guard forces were the main military presence within the New York City boroughs and on Long Island. Title 10 assets and personnel had been prepositioned at nearby Joint Base McGuire-Dix-Lakehurst (JBMDL), Trenton, NJ, and were awaiting mission assignments. However, growing frustrations over power outages, fuel shortages, and expanding news media coverage of the response, likely coupled with the pressures of the pending election, prompted the Federal Government to begin taking a more assertive stance in the response effort. The administration's October 31 "no red tape" guidance, coupled with NORTHCOM's November 2 fragmentary order (FRAGO),[4] while both certainly well-intentioned, contributed to some of the resulting confusion during subsequent days.

November 2 NORTHCOM FRAGO

- Get missions. Start with menu of DoD capabilities in the JOA that can be applied to support FEMA requirements.
- Do not wait for mission assignment paperwork. Coordinate with FEMA and the DCOs.
- Apply total force capabilities to accomplish missions. Operate on VOCO mission assignments when possible.
- When you get a mission: execute. Clean up paperwork later by coordinating with FEMA and the DCO.[5]

Those instructions, copied from a written FRAGO on November 2, seemingly encourages military commanders to abandon the structured processes normally in place in favor of less restrictive, verbal communication. In most cases, this demonstrates the military's ability to conduct flexible, adaptive, and in some cases, improvised operations when bureaucracy would simply be an unnecessary obstacle impeding efficiency. On its own, this guidance is encouraging and could be interpreted as consenting direction for Title 10 forces to respond under Immediate Response Authority. However flexible, it was in conflict with the DSCA process taught to military officers as part of their professional military education.

When guidance stems from the most senior levels, it tends to move through the subordinate echelons with greater urgency. As a result, actions often happen with more fervor and zeal. When such guidance contradicts policy and legislation, however, it invites violations of the same laws and procedures that are designed to maintain order, structure, and accountability in the first place. In some ways, senior commanders decided, essentially, to marginalize or ignore many of the guiding documents and laws governing military civil support operations. As a result, the National Response Framework (NRF), the mission assignment process, and other pertinent procedural guidance that serve as a system of checks and balances were largely ignored or abandoned by some senior leaders. The most notable example of this issue was the U.S. Marines landing on Staten Island without the prior consent or knowledge of the DSC.

Title 10 Awareness of the Dual Status Commander Construct.

Building on the earlier discussion detailing the events of November 4 and 5, we know that the Marines' arrival on Staten Island resulted from a series of conversations outside of the established chain of command and perhaps without consideration for normal Title 10 request for assistance procedures. A number of after-action reports and personal interviews with those knowledgeable of the events support the claim that the Commandant of the Marine Corps, through the II Marine Expeditionary Force commanding general directed the 26th Marine Expeditionary Unit (MEU) commander to deploy his unit to the USS *Wasp* off the coast of New York. The guidance from the Commandant instructed the MEU to: "Get to New York City, go ashore, do good, and relieve the suffering that is occurring."[6] These same reports and interviews suggest that a New York/New Jersey Port Authority official circumvented the normal processes and initiated the request via direct communication with Headquarters Marine Corps (HQMC). As a result, without a mission assignment or notifying the DSC, Marines carried out their orders and began support efforts on November 4. Except for justifying the Marine Corps' arrival on Staten Island as Immediate Response Authority (discussed in subsequent sections), the legal basis for the Marines' activity on Staten Island on November 4 and 5 is, at best, questionable. Since the DSC was unaware of the Marines' activity until after their arrival, this offers a valuable lesson learned to improve future Title 10 coordination with the DSC JTF. This failure of communication and coordination suggests several things worth discussing.

Incursion, intrusion, invasion, initiative: all are words that have been used to describe the Marines' landing on Staten Island on November 4. Aside from debating the semantic classification of the Marines' presence on Staten Island, this event illustrates some important points. Perhaps the most significant lesson learned from this action is the lack of familiarity and understanding of the DSC arrangement among some Title 10 officers. Some officers who commanded units during Sandy admitted to being completely unaware of the DSC concept, structure, and command arrangement prior to execution.[7] In addition, due to the pressure from the Commandant and the aforementioned NORTHCOM guidance to "get missions," the MEU repeatedly contacted the joint coordinating element at JBMDL, rather than the DSC JTF, to request mission assignments. This is problematic for two reasons. By contacting the joint coordinating element (JCE) and other senior commanders to request missions, the MEU:

- demonstrated that it did not have a clear understanding of the DSC chain of command and was, in effect, excluding the DSC from the conversation simply because it was unaware of the DSC role.

- was, in effect, pressuring senior commanders to involve a Title 10 force in the response effort prior to the DSC JTF's acknowledgement of the requested need.

After some time without acknowledgement from the JCE and following discussions with HQMC and Port Authority personnel, the MEU debarked a small detachment of Marines from the USS *Wasp* to assist Staten Island authorities in debris removal and restoration activities (a mission not covered by an approved mission assignment and without the awareness of the New York DSC at the time). Despite occurring outside of DSC's scope, the Marines provided a requested service in support of the residents and local authorities on Staten Island. As such, it became evident that this activity should continue. Following a brief interruption in activity on the evening of November 4 (as previously discussed), the Marines resumed support activities under the tactical control of the DSC on November 5.

The lessons learned here suggest several things. First, when command guidance encourages the abandonment of policies, accountability and clarity are lost in such a complex response environment. A long history of disaster research suggests that the "red tape" of government bureaucracy hinders response processes, often to the detriment of the citizens of an effected area.[8] As taxing as it may be to adhere to response policies and procedures, combined state and federal response efforts require some semblance of structure to function adequately. The Marines' landing on Staten Island suggests that not only do some military commanders lack the necessary education and knowledge concerning the DSC construct, but there is also limited understanding of the requirements and procedures of the mission assignment process. Regardless of the reason for abandoning procedures, this particular set of circumstances suggests that military commanders do not have a clear understanding of the dual status commander construct and its application during no-notice/limited-notice DSCA response scenarios. Moreover, between political influences, uncoordinated civilian activities, and occasional federal military ventures under Immediate Response Authority, DSC may not be able to command and control as much of the response as we expect them to. These events also suggest that the established command structure for the Sandy response was unclear.

Command Structure.

Beyond the Marine activity on Staten Island, confusion and lack of clarity concerning the actual joint command structure further complicated matters during the Sandy response. Because Sandy was a multistate incident, the decision was made to put a JCE as an intermediary echelon between the Army North commander (Lieutenant General William Caldwell) and the New York DSC. In this case, Major General Jeff Mathis, Commanding General of Joint Task Force-Civil Support served in this capacity as the JCE with supervision of all Title 10 forces in both New York and New Jersey during the Sandy response. In a single state incident, this would be an unnecessary command position, as the DSC would report directly to the Army North (ARNORTH) commander, to NORTHCOM, to the Secretary of Defense, and finally to the President on the Title 10 side. In the Sandy response, however, the JCE served as an additional command layer and added confusion to the already complex command hierarchy.

According to some accounts in after-action reports and personal interviews, the command structure changed multiple times during the first days of the operation. The role of the JCE was unclear to many, as there are conflicting accounts among those who participated in the event. Some maintain that the JCE was the intermediate link between the JFLCC (ARNORTH) and the DSC with command authority linking the two echelons (Figure 4-1). Others, however, dispute this, suggesting that the JCE's role was just that: a coordinating element with no command authority over the DSC, as suggested in the alternative structure shown in Figure 4-2.

While accounts differ, the fact remains that the command and control structure of the Sandy response was unclear to the Title 10 side. On more than one occasion, this lack of clarity resulted in the New York DSC fielding calls on his cell phone or receiving emails from Title 10 forces advocating for their capabilities and requesting orders to assist in the response. In effect, the DSC received multiple unsolicited requests from Title 10 forces petitioning for their inclusion in the operation. This not only points to a lack of clarity regarding the command structure, but also suggests that Title 10 forces either deliberately ignored processes or were mostly ignorant to the coordination and approval procedures involving the DCOs and their counterparts in the JFO. Within this context, other processes were equally challenging, leading to confusion during the Sandy response in New York.

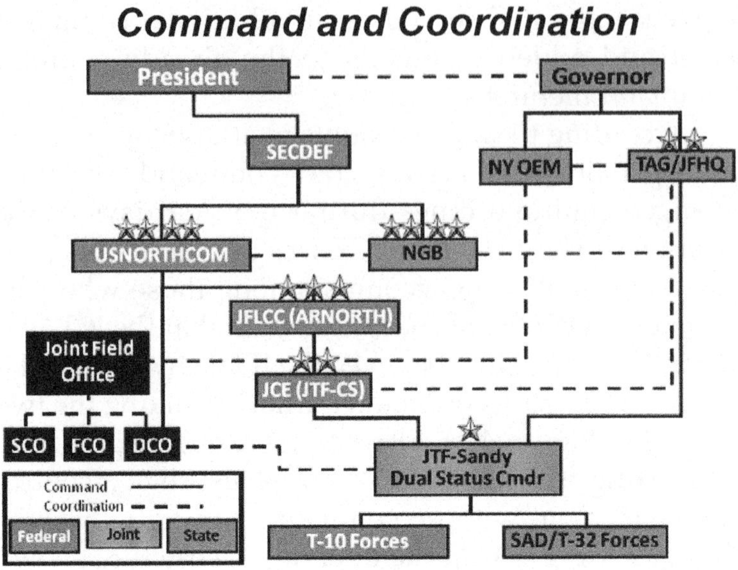

Figure 4-1. Hurricane Sandy Command Structure.

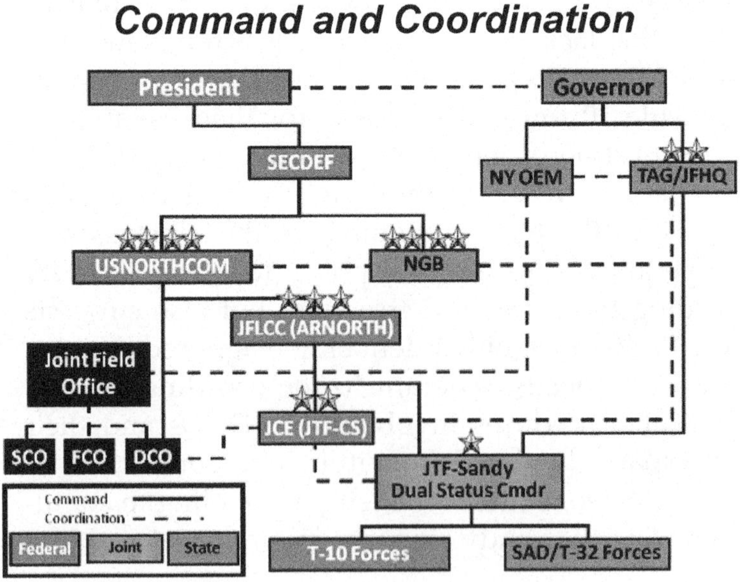

Figure 4-2. Hurricane Sandy Command Structure, Alternative View.

Mission Assignment Process.

The DoD mission assignment process outlining the procedures for Title 10 support of civil authorities is unwieldy. Combining this cumbersome process with the urgent needs following a no-notice/limited-notice incident creates additional burdens for military and civil authorities to manage. The unprecedented timing of Hurricane Sandy just prior to a presidential election and its near-direct hit on the most populated city in the United States only increased the interagency pressure to provide timely response. For reasons previously mentioned, the mission assignment process was not followed on several occasions during the federal response to Sandy in New York. The lack of adherence to established procedures can be attributed to all levels of command. Specifically, the Sandy response in New York suggests needed improvement in the mission assignment process as it relates to approval and authorization, as well as speed and necessity of assignment processing.

Within the mission assignment parameters, requests for DoD assistance are generated from local and state authorities after all other resources (local, county, state [including National Guard], and Emergency Management Assistance Compact) have been exhausted or are otherwise unable to provide the necessary service due to limited capabilities. (For example, dewatering the New York subways required pumping capabilities beyond local and state capabilities). Conceptually, this bottom-up process ensures that federal forces sourced for DSCA have a mission to perform based on requests from local authorities. In actuality, the mission assignment process sometimes creates a bureaucratic obstacle for a commander that hinders operational response. When there are unmet needs in a DSCA response scenario, bureaucracy creates frustrations and impatience, which can lead to noncompliance with established procedures. During Sandy, there were several instances of this as missions came from the top-down and without requests from state and/or local officials.

Without approved mission assignments but in consonance with NORTHCOM guidance, military units converged on the New York area in the days following Sandy's landfall. Units deployed intending to provide assistance without consideration for accounting procedures or Title 10 and Title 32 coordination strategies. This force surge created a cluttered JOA with some units in the area without the knowledge of the DSC. This also led to the inundation of unsolicited offers of support from Title 10 forces, which had to be fielded by members of the JTF staff, further complicating an already complex coordination effort. In some instances—and likely due to the sluggish mission assignment approval process—missions were generated and disseminated from the top-down, rather than bottom-up, per the NRF guidelines. Eventually, Title 10 forces began conducting support activities without a mission assignment or knowledge of the DSC. As a result, key personnel in the JTF staff, the New York State Office of Emergency Management, including the state coordinating officer (SCO), and other critical coordination elements, were excluded from the conversation. This led to increased confusion and reduced interagency coordination. With Title 10 forces operating in the DCS's tactical area of control and responsibility and without a valid mission assignment, the immediate reaction in some instances was to order all nonapproved activities halted until approval was granted. As a resource employer, the DSC is responsible for coordinating federal and state efforts simultaneously. Mission assignment protocol should not restrict the DSC from employing state and/or federal resources appropriately to meet a need. The impediments that prevent a DSC from approving missions within his/her area of responsibility, especially Title 10 missions providing needed support to local authorities, hence, should be removed.

To many, the published mission assignment process is mechanistic and convoluted. Some argue that the heavily bureaucratic process creates delays and inefficiencies at a time when speed and effectiveness are most critical. Therefore, the fact that DoD did not adhere to the mission assignment process during the Sandy response may be perceived by many as a progressive step forward. However, abandoning the mission assignment process entirely creates significant impediments to coordinating and executing an operationally, legally, and financially sound federal response. With increased confusion resulting from ad hoc processes outside of the established guidelines, this ultimately diminishes the unity of effort desired in DSC-led DSCA responses.

As with other topics discussed, the Sandy response in New York illustrated some of the current issues plaguing the process and offers a useful case study platform to generate improvement. Based on this event, it is clear that the mission assignment process can be improved to ensure this situation does not occur again in future DSCA missions. During a no-notice/limited-notice incident, the first 72-96 hours of the federal response are absolutely critical and can mean the difference between a manageable disaster and one where Congress, the media, and the public collectively blame DoD for sluggishness. A better and more streamlined mission assignment process is needed to recognize this critical response period. However, improving the mission assignment process is only one step in the larger DSCA context. As the response to Sandy demonstrated, some military commanders and other senior defense officials lack the requisite knowledge of the DSCA environment and the newly established DSC construct. We must ensure that senior military commanders and defense officials are fully educated in the DSCA arena, with specific emphasis on the DSC arrangement.

Defense Support of Civil Authorities Education.

Of the noted areas needing improvement, perhaps none is more important than DSCA education for senior military leaders. While there are many subject matter experts in all things related to defense support of civil authorities, there appears to be a critical gap in DSCA knowledge among some senior military commanders. As evidenced by the failure to follow mission assignment processes and the notable confusion over the role and authority of the DSC, it appears that some senior leaders, often with decisionmaking authority, lack the required knowledge to ensure their decisions fall within established legal, financial, and doctrinal barriers of DSCA operations. The critical triad of DSCA considerations—the legal, financial, and doctrinal guidelines—were abused during the Sandy response in New York, in many cases due to a lack of DSCA knowledge among commanders and their support staffs. This is not to suggest that all Title 10 DSCA operations were in violation of policies and procedures; much to the contrary, in fact. The problems that occurred during the Sandy response in most cases can be attributed largely to lack of formal education, training, and knowledge of the DSCA response environment.

As noted previously, some Title 10 force commanders were unaware of the DSC arrangement prior to their involvement in Sandy. The same reports and interviews suggest that USACE commanders were equally unfamiliar with the DSC construct. In this case, Title 10 forces attached to Task Force Pump and in support of the USACE as the lead federal agency for Emergency Support Function-3 were assigned missions beyond the scope of any pre-approved mission assignments for Title 10 forces. Reports suggest that USACE personnel were unaware of certain Title 10 restrictions for federal military forces and did not have an effective process in place to facilitate coordination with the DSC. While these issues were resolved through effective interservice liaison efforts, they point to larger issues that must be addressed.

If unit commanders supporting DSCA operations are unaware of the command structure in place, this can certainly contribute to increased confusion and uncertainty, much like what occurred during Sandy. The DSC construct is relatively new and had never been used during a no-notice/limited-notice DSCA response prior to Sandy, so there is some expectation of unfamiliarity. However, the lack of knowledge toward the DSC arrangement on behalf of some commanders during the execution of a real-world DSCA operation is troubling. This suggests that we need to improve knowledge and awareness of senior military officers with DSCA-related mission capabilities organic to their units. It also suggests that we need to significantly improve our communication and information sharing prior to and during DSCA operations so that commanders supporting civil authorities can operate within the established command configuration, limit future confusion, and therefore, contribute to the desired unity of effort that the DSC arrangement is designed to facilitate. Finally, this suggests an urgent need for more realistic training and exercises like Vigilant Guard, Ardent Sentry, and others designed to simulate a DSCA response under a dual status commander.

Dual Status Commander Guidance/Instructions.

Much of the confusion and lack of situational awareness concerning the DSC initiative can be attributed to the lack of formal guidance currently contained in DoD reference publications, doctrine, and instructions. Currently, the DoD Instruction 3025.xx, "Dual Status Commanders for Defense Support of Civil Authorities," is in draft status, with a tentative fall 2014 release. As a subordinate publication to the more widely circulated DoD Directive 3025.18, "Defense Support of Civil Authorities,"[9] DOD Instruction 3025.xx will address many of the current issues of confusion concerning dual status commander-led DSCA operations. Until the release of this new instruction, few other defense references offer guidance on the dual status commander construct, and even fewer contain any substantive information that can be applied to no-notice/limited-notice incidents.[10] There is a critical need within DoD and the National Guard to codify DSC guidance through the development and continued revisions of relevant guidance, doctrine, and reference publications. Such work is ongoing within the Office of the Secretary of Defense, NORTHCOM, and the individual services. However, until DoD develops and releases clear, well-articulated guidance detailing the many issues relevant to the DSC construct, these operations will continue to experience challenges like those noted.

While not a comprehensive list, this chapter offers a brief description of some notable areas recognized through the Hurricane Sandy operations as needing improvement during dual status commander-led DSCA operations. Using this as a basis for future decisionmaking and planning efforts can lead to improvements in these critical mission capabilities under the unique command arrangement noted.

ENDNOTES - CHAPTER 4

1. Interviews conducted by the author with various DoD employees, National Guard officers, and active military officers that participated in the response to Hurricane Sandy in some capacity, January-March 2014.

2. *Ibid.*

3. "Don't be late" was a repeated axiom describing DoD's strategic planning approach to the Sandy response. This was noted in multiple after-action reports and in the context of the lessons learned from Hurricane Katrina.

4. "Fragmentary Order" is defined by *Joint Publication (JP) 1-02, Department of Defense Dictionary of Associated Military Terms,* as "An abbreviated form of an operation order issued as needed after an operation order to change or modify that order or to execute a branch or sequel to that order." JP 1-02, Washington, DC: DoD, 2014, p. 105.

5. Copied material from NORTHCOM guidance (fragmentary order) of November 2.

6. Summary of guidance issued by the Commandant of the Marine Corps provided to the authors through personal interviews and multiple DoD after-action reports detailing the circumstances of the Marines' response.

7. Various DoD after-action reports suggest Title 10 unit commanders were unaware of the DSC construct or how federal military forces integrate into the joint structure.

8. David Neal and Brenda Phillips, "Effective Emergency Management: Reconsidering the Bureaucratic Approach," *Disasters,* Vol. 19, Issue 4, 1995, pp. 327-337; Gary Webb and François Chevreau, "Planning to Improvise: The Importance of Creativity and Flexibility in Crisis Response," *International Journal of Emergency Management,* Vol. 3, Issue 1, 2006, pp. 66-72; Gregory Streib and William Waugh, "Collaboration and Leadership for Effective Emergency Management," *Public Administration Review,* Special Issue, December 2006, pp. 131-140; William Waugh, "Mechanisms for Collaboration in Emergency Management: ICS [Incident Command System], NIMS, and the Problem with Command and Control," Rosemary O'Leary and Lisa Blomgren Bingham, eds., *The Collaborative Public Manager: New Ideas for the Twenty-First Century,* Washington, DC: Georgetown University Press, pp. 157-175; Russell Dynes, "Community Emergency Planning: False Assumptions and Inappropriate Analogies," *International Journal of Mass Emergencies and Disasters,* Vol. 12, No. 2, 1994, pp. 141-158.

9. DoD Directive 3025.18, *Defense Support of Civil Authorities,* was last revised in September 2012 and contains no language addressing the DSC construct in any capacity.

10. The only DoD publication devoted to the DSC construct at this time is *NORTHCOM Publication (NP) 3-20, Title 10 Support to Dual Status Command Led Joint Task Force Standard Operating Procedures*. Released in January 2012, this document outlines the employment procedures and considerations for the use of DSCs during civil support missions. However, this document pre-dates Hurricane Sandy by nearly a year. Lessons learned from Sandy are beginning to trickle in and have led to the need to rewrite this publication. According to NORTHCOM personnel, NP 3-20 is undergoing significant revisions at this time. JP 3-28 was published in July 2013 following significant revisions from the 2007

version. JP 3-28 offers some of the most comprehensive text regarding DSC of all DoD publications reviewed. This new version "introduces, defines and clarifies the dual-status commander to include nomination, training and appointment requirements" (p. iii). Additionally, JP 3-28 includes a useful process diagram (Figure 13) to depict the DSC designation process once requested by state governors (2013, p. C-9). *U.S. Army Field Manual (FM) 3-28, Civil Support Operations* provides a detailed description of the DSC concept and construct. This reference defines the authorities and requirements for establishing DSC arrangements and provides useful graphics to illustrate the operational and tactical command relationships between the DSC, State Governments, and the Federal Government (p. 7-5). However, defined guidance on the execution process for no-notice/limited-notice incidents is absent in this reference. Other pertinent military reference publications, including *Army Doctrinal Publication 3-28* and *Multi-Service Tactics, Techniques, and Procedures (MTTP) 3-28,* address the DSC concept briefly without providing any substantive guidance for the execution of complex no-notice/limited notice DSCA operations such as Hurricane Sandy.

CHAPTER 5

RECOMMENDATIONS FOR IMPROVEMENT

Building on the previously stated issues and the preceding analysis of the defense support of civil authorities (DSCA) response to Hurricane Sandy, this final chapter of the monograph outlines a detailed series of strategy and policy-related recommendations specific to no-notice/limited-notice DSCA responses under a dual status commander (DSC)-led joint task force (JTF). The following recommended actions and measures are intended for senior Department of Defense (DoD) leaders to consider to improve potential DSC improve the DSC construct and related concepts within defense support of civil authorities operations.

The suggestions are based on an extensive analytical coding process that identified recurring themes in the source data (interviews, after-action reports, etc.). After coding and interpreting the material to identify viable recommendations, we again coded the material into two distinct categories using an axial coding approach. The recommendations are grouped into two categories: 1) operational and strategy-specific, and 2) policy-specific. A brief narrative description or justification accompanies each recommendation. The material offers a condensed summary of the most advocated recommendations from throughout DoD. The recommendations are intended to be actionable and realistic; although some, if implemented, require significant changes or alterations to existing policies, procedures, or doctrine, and may, therefore, be judged impractical by some.

OPERATIONAL AND STRATEGIC RECOMMENDATIONS

Lean, But Don't Push Forward.

The forward-leaning approach employed by DoD prior to and during the Sandy response was effective. Prepositioning Title 10 forces at nearby bases and offshore, issuing prepare-to-deploy orders, and deploying defense coordinating officers to anticipated disaster areas is necessary to facilitate a timely response upon request from civil authorities. Cost issues aside, moving Title 10 forces into the joint operations area (JOA) provides the DSC with abundant force capabilities ready to meet nearly every contingency. Some officers who participated in Sandy suggested that prepositioning Title 10 forces was the preferred strategy, rather than activating National Guard troops through Emergency Management Assistance Compact (EMAC) and other sourcing mechanisms.[1] Aggressive posturing of Title 10 forces in and around a disaster area offers quick response and unmatched capabilities. However, DoD and the service components must avoid being too forceful while ensuring compliance with laws, policies, and procedures.

Title 10 forces and commanders should be encouraged to comply with national guidance. The concern over a forward-leaning approach arises when Title 10 forces are not integrated into the response as expected following deployment to the affected area. Federal funds are used to transport units to the JOA. Upon arrival, commanders often search for opportunities to integrate their forces into the DSCA response to justify the cost of transport, among other things. This external pressure can have a detrimental effect and should be avoided to the extent possible. Therefore, DoD should continue to preposition assets and personnel when there is an anticipated need. However, despite public and Hollywood mythology, federal military forces are not the nation's principal emergency response service. Therefore, upon arrival, commanders should refrain from asking for Title 10 integration and instead wait until they are requested through the established methods.

Delineate Clear Federal Chain of Command Prior to Deploying Forces.

To avoid similar confusion regarding the chain of command structure in place for Sandy, once the decision is made to activate a dual status commander for a joint DSCA response, both DoD and the affected state(s) should clearly articulate and approve a unified chain of command. The roles, responsibilities, and lines of command and coordination, respectively, must be clearly established prior to the deployment of a joint task force. During Sandy, there was confusion over the role of the joint coordinating element (JCE) and whether the JCE was internal or external to the federal chain of command. As previously addressed, some commanders who participated in the Sandy response contend that the JCE was the parent command element to the DSC and therefore served as the command link between the DSC and the joint force land component commander (JFLCC)-Army North (ARNORTH). Conversely, others claim the JCE was simply a coordination element with no command authority over the DSC. In this view, the DSC reported directly to the JFLCC on the federal side. This confusion and lack of clarity among participants during Sandy created additional challenges that should be avoided in future DSCA operations. As early as possible, U.S. Northern Command (NORTHCOM) should clearly articulate a federal chain of command, including the names and titles of each command echelon down to the DSCs. Command and control wire charts should be created and disseminated prior to operations, to the greatest extent possible given the circumstances.

Eliminate the Joint Coordinating Element.

According to senior DoD officials, the inclusion of a JCE during the Sandy response was a trial concept intended to improve coordination efforts between multiple dual status commanders and the federal chain of command during a multistate incident.[2] Due to the increased confusion presented by the inclusion of a JCE during Sandy, this command element should be removed from future consideration. Some advocate for inclusion of a JCE during a DSCA response to facilitate Title 10 force coordination, including joint reception, staging, onward movement, and integration (JRSOI). For a multistate event such as Sandy, effectively coordinating Title 10 force activities logically warrants consideration of a JCE. The problem occurs when the JCE commander's role is not clearly articulated. If using a JCE for the purpose of effective JRSOI of Title 10 forces, the JCE should not be included as part of the formal command structure. Instead, the JCE should be listed as a coordination entity (dotted line doctrinally) only. If a JCE is not desired for JRSOI, then removing the JCE entirely from the federal chain of command reduces the layering effect noted during the Sandy response. Without a JCE, the dual status commander can and should report directly to the JFLCC/ARNORTH commander as the parent command entity. Figure 5-1 illustrates how this command structure would have looked during Hurricane Sandy.

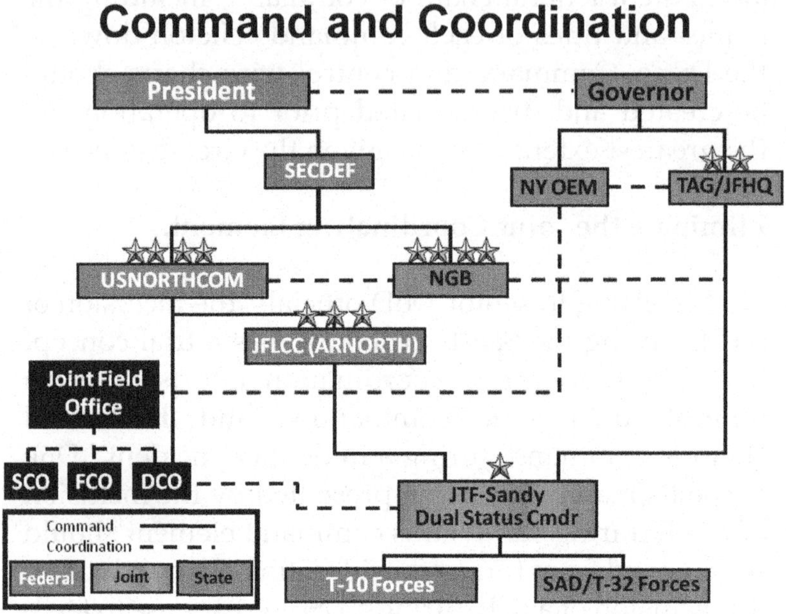

Figure 5-1. Hurricane Sandy Command Structure, No JCE.

Appoint a Defense Coordinating Officer in Charge.

National response plans, such as the National Response Framework (NRF) and other guiding documents, call for one DCO to serve as the single point of contact at the joint field office (JFO) for DoD activities within each of the 10 Federal Emergency Management Agency (FEMA) regions. However, as noted in *Joint Publication (JP) 3-28*, larger incidents and multistate incidents sometimes require additional DCOs to assist in coordination efforts.[3] According to DoD after-action reports and interviews, multiple DCOs deployed to New York during Sandy, each with a defined area of responsibility but no guidance for DCO-DCO coordination.[4] Similar to the JCE, deploying multiple DCOs creates the potential for confusion and coordination issues. There were several issues noted in Sandy after-action reports detailing failures of DCO-DCO coordination within New York, resulting in duplicated planning efforts or redundant mission assignment generation—something that the DSC concept is designed to help alleviate.

Currently, there is no defined adjudication process between the multiple DCOs assigned to a FEMA region and NORTHCOM. In essence, NORTHCOM may receive duplicate requests from different DCOs in the same region who have no established protocol to coordinate with each other. The suggested defense coordinating officer in charge (DCOIC) billet will serve as this needed adjudication body for all DCO activity during designated incidents involving multiple DCOs. In this model, DCOs will preliminarily approve requests for forces and submit them to the DCOIC for final approval. The DCOIC will validate or deny these requests and inform the DSC accordingly.

There are multiple options for designating a DCOIC. One alternative is to simply designate a DCO as the senior DCO in charge based on established criteria (rank seniority, time in billet, etc.). In this model, all subordinate DCO activities are routed through the DCOIC to ensure effective coordination, reduced redundancy, and a unified mission assignment process. The DCOIC would coordinate directly with the DSC JTF as the DoD representative to the JFO. An additional alternative to this recommendation places a general officer in the DCOIC billet. In this model, the

Joint Task Force-Civil Support (JTF-CS) commanding general (CG) is an ideal candidate for the DCOIC position, as this person is one of the leading subject matter experts on domestic civil support and DSCA. Assuming a temporary assignment like DCOIC does not conflict with the JTF-CS CG's principal chemical, biological, radiological, or nuclear DSCA responsibilities, this general officer should hold a key role in the execution of DSCA operations, especially those using the DSC arrangement. This recommendation also complements the previous suggestion to remove the JCE from the command structure. Rather than assigning the JTF-CS CG to serve in a billet of questionable necessity (JCE), this general officer can be deployed to the designated JFO to serve as the DCOIC with ultimate approval authority over all DCO-authorized mission assignments in the DSC JOA.

Regardless of the chosen option, establishing an adjudication body for the multiple DCO constructs likely to occur again in major incidents will help limit future confusion and redundancies.

Define Time for Early Title 10 Integration.

Hurricane Sandy gave federal military forces the opportunity to highlight their timely response capabilities. As members of a professional military force, Active Duty Soldiers, Sailors, Airmen, and Marines remain in a constant state of readiness to deploy and respond to domestic contingency operations. This quick response capability was displayed during Sandy as various Title 10 assets converged on the JOA within days of the storm's landfall. The Active Component's ability to mobilize and deploy forces quickly, coupled with the Reserve component's widespread geographic distribution of forces, often means Title 10 forces are able to respond quicker and with more capabilities than their National Guard counterparts. However, Title 10 forces come with a greater financial requirement. Therefore, through state National Guard capabilities and EMAC agreements, National Guard assets are the primary military sourcing solution for disaster response operations. Whether due to administrative delays through EMAC requests or insufficient capabilities, the National Guard cannot always address immediate requirements, such as the need for dewatering the New York City subway system during Sandy. In these cases, states look to the Federal Government for support. With this in mind, DoD and the governors should consider a strategic shift that would allow federal forces to be sourced, following a request from civil authorities and external to Immediate Response Authority (IRA), for a predetermined (and finite) period of time during the early phases of a DSCA response. This can be done prior to sourcing National Guard units during the initial stages of DSCA operations to facilitate quicker military response when necessary.

The proposed Title 10 integration period should not be misconstrued as an unrestricted authorization for the DSC to employ Title 10 forces. Rather, this should be considered a defined period of time when Title 10 forces can be sourced prior to the National Guard's arrival and without an approved mission assignment. In many ways, this is similar to IRA in that the proposal allows federal forces to provide assistance without the need for paperwork delays. However, this is different from IRA in that the DSC approves the Title 10 integration and therefore assumes tactical control of the federal force. The DSC does not have tactical control of Title 10 forces operating under IRA, per JP 3-28.[5] Similar to the policy governing IRA, this Title 10 integration period should extend at least 72 hours from the activation of the DSC. During this period and assuming consent of the federal and state commands respectively, the DSC should be authorized to use Title 10 forces to address priority requests for assistance with consideration of cost share and mission assignment generation after the fact.[6] The integration period should also establish a clear time limit (e.g., 96 hours) to conclude all initial Title 10 sourcing. At the conclusion of the proposed integration period, the DSC can prioritize National Guard forces for secondary and tertiary requests. This will facilitate flexibility and adaptation to the evolving situation and limit the bureaucratic delays present in the current system.

The proposed Title 10 integration period will provide a mechanism to address external pressures to involve Title 10 forces in DSCA operations. Ideally, Title 10 forces would exercise better fidelity to the "last in, first out" philosophy during domestic response. However, the political realities of domestic response often supplant policy and law. The President, governor, and other elected representatives risk a political death sentence for inadequate, insufficient, or late response in events of national prominence or significance. As evidenced by the federal response to Hurricane Sandy, politicians will often marginalize or abandon restrictive laws and policies in order to provide immediate federal assistance and avoid public ridicule. Providing a mechanism to integrate Title 10 forces into the early stages of a DSCA operation will address these concerns while ensuring prompt assistance to civil authorities when requested. It will further reduce the tensions over Title 10 activities under IRA, as federal forces will be integrated into the response and under the tactical control of the DSC.

Simply put, if DoD and the states are going to embrace the DSC concept, per the 2012 National Defense Authorization Act, as the usual and customary command arrangement during the simultaneous employment of the National Guard and Armed Forces, we must empower the DSC in such a way that leverages all available resources and capabilities, both state and federal. The DSC should be able to request Title 10 forces to meet a need within the JOA so long as such support is not illegal, immoral, or unethical. There are too many impediments in the current process restricting commanders from providing the best and most capable response resources in a timely manner, while also encouraging the abuse of less restrictive policies like IRA to justify response activities. This mechanism will help address some of the noted issues.

Authorize Transition of Authority.

While the previously discussed recommendation addresses initial Title 10 activities requested by the DSC, it does not address similar activities performed outside of the DSC's knowledge under IRA. DoD Directive (DoDD) 3025.18 provides commanders IRA when requested by a civil authority and under "imminently serious conditions and if time does not permit approval from a higher authority."[7] In these instances and where Title 10 forces are operating within the DSC's joint operations area, consideration should be given to whether the DSC should assume tactical control of federal forces operating under IRA. This topic is debated regularly in and around DoD, with perspectives advocating both for and against such a recommendation.

Those who support the DSC gaining tactical control of Title 10 forces under IRA cite joint doctrinal concepts, such as unified action and unity of effort[8] as a basis for their argument, often noting that when Title 10 forces operate under IRA and thus outside of the DSC JTF, neither is achieved. Instead, their argument holds that any force operating outside of the command of the DSC JTF creates friction rather than promoting synergy and unity of effort. Advocates further contend that one of the main intentions for creating a DSC architecture is to unify state and federal military actions under a single commander, albeit in a mutually exclusive capacity. Title 10 forces operating externally to this joint command structure are not in consonance with the unified, coordinated concept the DSC is designed to facilitate.

The counterargument to this position is rooted in the tenets of federalism and the division of powers between the States and Federal Government. As noted earlier, the legal framework guiding the use of military forces domestically is complex. Despite the complexities, however, critics affirm the constitutional basis of the laws and philosophical principles as the foundational structure for using the military domestically. According to the Constitution—and with support from Title 10 of the United States Code—the President is the commander in chief of the Armed Forces under all circumstances. When responding under IRA, Title 10 forces maintain autonomy from the DSC, instead reporting directly to their service commanders and, in effect, to the President. Permitting a DSC (who in most cases is a National Guard officer) to assume tactical control of a Title 10 force under IRA, according to critics, contradicts not only the doctrinal restrictions prohibiting DSCs from exerting command authority over Title 10 forces under IRA,[9] but also the principles of federalism that are intended to ensure a divided system of power and authority between the states and national government. Using this logic, opponents even suggest that the mere concept of a DSC violates the Constitution and the federalist system of government.[10]

Regardless of perspective, DoD must determine a policy and strategy for coordinating with or integrating Title 10 forces on IRA during a DSC-led DSCA incident. While there is some question as to whether the Marines were operating under IRA during their initial arrival on Staten Island, the presence of a Title 10 force ashore during Sandy without the knowledge of the DSC created avoidable tension and confusion. In this case, the DSC in New York assumed tactical control of the Marine detachment ashore on Staten Island, following a series of discussions with other general officers within the chain of command. This tactical control ceased once the Marines returned to the ship. Despite initial disagreements, the assumption of tactical control of the Marines worked under these circumstances. Once the DSC gained situational awareness of the Title 10 activities ashore, he was better able to integrate their capabilities into future missions and support activities. If this is determined to be the most desirable course of action for future incidents of similar circumstances, there should be a process or procedure in place for the DSC to assume tactical control of Title 10 forces under IRA without having to go through several layers of command discussions. Defining such procedures in future doctrinal references will help future DSCs avoid the lengthy command discussions and negotiations that occurred during Sandy.

Designate and Employ a Title 10 Adaptive Task Force.

Much of the Title 10 activity during the Sandy response occurred under Task Force (TF) Pump, a joint force represented by all four services and responsible for numerous dewatering missions throughout the JOA in New York. This model worked well during Sandy. As a somewhat ad hoc and hastily requested force asset, TF Pump gave the DSC the tactical flexibility to employ Title 10 forces for specific missions related to dewatering, pumping, etc. TF Pump received most of the requests for dewatering and subsequent mission assignments falling under this special capability. This provided the DSC with at least one clear decision point during the entirety of the response operation. Similar actions should be considered for future missions.

Given the notable successes of TF Pump during Sandy, DSCs, in consonance with their state and federal chains of command, should identify a large critical mission capability (such as dewatering during Sandy) during the initial stages of a response effort. After agreeing on this capability requirement, NORTHCOM should identify and designate a unit capable of providing such services. This unit should be issued prepare-to-deploy orders and assume the designation as the Title 10 adaptive task force. Once identified, the DSC can exercise the option to activate the adaptive task force to complete mission assignments within the task force's identified specialty. Predetermining a Title 10 task force for performing specific mission functions will limit the tensions between Title 10 and Title 32 commanders lobbying for inclusion of their respective assets.

As with the other recommendations, there are counterarguments to this as well. It is difficult to predict future incident requirements, so employing an adaptive task force may not always be a possibility. Beyond this, there is a cost element associated with a federal military force supporting civil authorities. Financial considerations are (or should be) external to the DSC's focus during a DSCA response. However, cost is something that must be considered when determining whether to deploy any Title 10 force in support of civil authorities. With this in mind, some might suggest that a Title 10 task force represents unnecessary redundancy that can otherwise be sourced from existing unit capabilities. While these are valid considerations, designating an adaptive Title 10 task force is still worth considering, based on the observations from TF Pump during Hurricane Sandy.

Maximize the Use, Distribution, and Presence of Liaison Officers.

> You can't have enough LNOs in my opinion.[11]
>
> National Guard Interview

One of the most frequently discussed topics following the Sandy response was the use of liaison officers (LNOs) throughout the operation. Numerous interviews and after-action reports noted the importance of using LNOs to coordinate efforts and enhance situational awareness across the seemingly endless bureaucracy of local, state, and federal agencies, departments, offices, and services participating in the response. The mostly positive feedback concerning LNOs suggests that these positions are vital to successful coordination and information sharing of future DSCA response efforts, especially under the DSC construct when both states and the Federal Government are represented.

LNOs provided critical information to commanders and their staffs during the entire Sandy response effort. While they were used in many places, some agencies or offices did not have LNOs representing all relevant military units. In addition to providing a Title 10 and Title 32 LNO to every major command element, including both the DSC and any adaptive Title 10 task force, LNOs should also be located in the State Office of Emergency Management or Emergency Operations Center, and with FEMA to facilitate mission assignment coordination and subsequent force packaging. Additionally, some LNOs were underutilized according to various reports. Given the critical capability and knowledge provided by LNOs, assigning liaison personnel to perform staff functions is not an effective way to leverage their presence as subject matter experts in coordination. DoD and the National Guard should continue using LNOs in every location deemed necessary and ensure they are used in a manner consistent with their capability and expertise.

POLICY RECOMMENDATIONS

The most needed recommendation is to establish a DSC policy. Short of that, there are several policy revisions worth considering to improve future DSCA efforts involving the DSC arrangement. The most pressing change, based on observations from Hurricane Sandy, centers on the IRA provision in DoDD 3025.18. Additionally, revisions need to be made to the mission assignment process and certain Title 10 and Title 32 legislation, among other things.

Immediate Response Authority.

Revise and Codify Definition of Civil Authority.

The current language describing IRA in DoDD 3025.18, 4(g) states:

> In response to a request for assistance from a civil authority, under imminently serious conditions and if time does not permit approval from higher authority, DoD officials may provide an immediate response by temporarily employing the resources under their control, subject to any supplemental direction provided by higher headquarters, to save lives, prevent human suffering, or mitigate great property damage within the United States.[12]

The policy language here, according to DoD officials, is intentionally vague to allow for flexibility in the interpretation of what constitutes a civil authority.[13] The intended ambiguity allows for open interpretation based on individual circumstances and provides justification for military commanders to offer critical support to civil authorities without the need to subject their decisions to a lengthy and often-cumbersome approval process. This ensures that when American citizens have an immediate need for military support, the language of a policy does not prevent saving lives and alleviating suffering. Conceptually, this is sound logic based on the best interests of the American people. In practice, however, there are issues with the current wording that can lead to abuse of the provision or arguments over whether federal military action actually constitutes IRA.

Among the many ambiguities, the IRA policy fails to define an appropriate level of civil authority to request assistance from DoD using the IRA justification. As noted earlier, U.S. Marines from the 26th Marine Expeditionary Unit (MEU) came ashore on Staten Island at the request of an unidentified employee of the New York/New Jersey Port Authority.[14] While these actions have not been formally designated as IRA, as some question whether "imminently serious conditions" were present, there is no other legal basis for justifying the Marines' presence on Staten Island during the initial response period. Based on previous language, "the Marine invasion of Staten Island," as it has been referred to, was, in effect, compliant with at least part of Section 4(g) of DoDD 3025.18, in response to a request for assistance from a civil authority. Aside from revisiting the semantics and meaning of "under imminently serious conditions," DoD should consider revising, expanding, and clarifying the description of "civil authority" as it applies to IRA.

In its current form, the term "civil authority" is ambiguous and leaves significant room for interpretation. According to JP 1-02, *Department of Defense Dictionary of Military and Associated Terms*, civil authorities are:

> Those elected and appointed officers and employees who constitute the government of the United States, the governments of the 50 states, the District of Columbia, the Commonwealth of Puerto Rico, United States territories, and political subdivisions thereof.[15]

This broad definition lacks the specificity needed for DSCA operations and, legally speaking, facilitates federal military actions in response to a request from any level of civil authority without restriction. Others can debate the necessity of the Marines' presence on Staten Island and whether "imminently serious conditions" existed. The emphasis here is that DoD should consider revising the term civil authority(ies) to avoid future confusion and/or abuse of the IRA policy. The revision should specify distinct levels of civil authorities on a hierarchical scale; or, where appropriate, titles of positions. It should further designate what level constitutes an appropriate requesting authority (e.g., "In response to a request from a Level 3 civil authority . . ."). Such policy revisions are needed to avoid similar problems in future response efforts. In addition to expanding on the appropriate level of civil authority to which the DoD can respond under IRA, the provision should be revised to expand and clarify "supplemental direction."

Supplemental Direction.

The IRA guidance permits military commanders to engage in immediate response "subject to any supplemental direction provided by higher headquarters."[16] In order to avoid abuse of IRA in future DSC-led DSCA efforts, DoD should create a standardized "supplemental direction" for reference under IRA. In this context, DoD and/or NORTHCOM should consider drafting a template standing order or directive to augment or serve in addition to the current Chairman of the Joint Chiefs of Staff DSCA Standing Execution Order. This additional standing order shall be applicable specifically to DSC-led responses and instances of IRA under which federal forces may operate. This order should be issued by the NORTHCOM CG and hold all Title 10 force commanders accountable to a specified standard of conduct or procedure when providing assistance under IRA.

For example: U.S. Marines of all ranks are intimately familiar with the 11 general orders of a sentry. Just as a marine can receive punitive action for quitting his/her post without being properly relieved (general order 5), a military commander should be deterred from abusing or violating federal policy on DSCA. Similar in style and custom to the general orders of a sentry, NORTHCOM should develop a standing general order to guide the conduct of DSCA operations under IRA, with emphasis on those occurring within the DSC construct.

Example General Order for Immediate Response Authority within a DSC JOA: Commanders using Immediate Response Authority to support civil authorities within a dual status commander joint operations area must notify the dual status commander within three hours of authorizing immediate response.

Currently, there is no incentive for commanders to ensure the integrity of DSCA doctrine, policies, and procedures. While the legal basis for some Title 10 actions during Sandy is questionable, it is nearly inconceivable to think of a situation where such violations of policy or law would result in punitive action against the responsible commander, nor are we suggesting that commanders should be punished for providing immediate response, especially when the actions of response forces are carried out with the intent to provide assistance to local residents. However, without changes or additions to the current policy, there is ongoing potential for similar issues in future DSCA activities. Issuing a combatant command endorsed general order prior to the execution of a DSCA response would provide the necessary mechanism or incentive for command compliance with standing laws, policies, and procedures. The revision to the IRA guidance could read: DoD officials may provide an immediate response by temporarily employing the resources under their control, in accordance with NORTHCOM general order or any additional supplemental direction provided by higher headquarters.

As with other recommendations, this suggestion may be unpopular with some. While the need to maintain speed and flexibility during DSCA is critical, maintaining accountability and awareness of response activities is important as well. In many cases, creating additional layers of policy compliance slows response decisionmaking and operational efficiency. In this case, however, requiring a single notification from a Title 10 commander to the DSC does not add to existing restrictions, nor does it limit a commander's ability to provide support under IRA. This recommendation simply ensures Title 10 commanders exercising IRA provide the DSC with appropriate notification of their intent and ongoing activities up to the 72-hour period of authorization. This contributes to the goal of achieving unified actions and an overall unity of effort under a DSC-led JTF.

Mission Assignment Process.

The recommendations given earlier offer a mechanism for clarifying preliminary considerations and approval measures for Title 10 actions under the pretext of IRA. However, the DSC in New York encountered several issues after Title 10 forces arrived, most of which can be improved through changes to current mission assignment policies.

Assuming Title 10 actions meet all established criteria for IRA, the DSC may wish to sustain this support activity beyond the currently approved 72-hour authorization period. To facilitate sustained Title 10 activity beyond the first 72 hours of IRA, a mission assignment must be generated and approved through the appropriate channels, or Title 10 forces risk being subjected to a work stoppage request, as was the case with the Marines on Staten Island. To avoid such problems, DoD, in conjunction with the requesting local/state agency, should codify a process by which a mission assignment/formal request for assistance is generated and submitted through the proper channels. Developing a post-IRA mission assignment/request for assistance process and incorporating it into current policies will:

- Fill a current gap in which no policy guidance exists for actions occurring beyond the initial 72-hour period under IRA.
- Provide a policy/doctrinal basis for DSCs to assume tactical control of Title 10 forces operating within the JOA, if desired.
- Provide a needed policy mechanism for reimbursement of Title 10 support activities that will eventually fall under an approved mission assignment.

Without the restrictions, civil authorities with knowledge of the IRA policy language can ignore the current mission assignment and/or request for assistance process while leveraging the ambiguous language of IRA. As seen during Sandy, this can lead to violations, intentional or not, of policy and, in some cases, law. The absence of this essential guidance further marginalizes the essential considerations for initiating requests for DoD support, often leading to greater end costs and confusion.

With this in mind, DoD should consider developing a draft instruction outlining the specifics of the mission assignment process, to include when and how Title 10 forces operate under IRA.[17] The political pressures and realities of a response situation can cause the established system of accountability to be circumvented, or in some cases abandoned. Sending troops "towards the sounds of chaos"[18] may be politically convenient for elected officials seeking public approval and for military commanders cleverly seeking a boost in their service's recruiting mission and budget appropriations, but it can also impede and aggravate planned and coordinated response efforts.

Final Thoughts.

Again, in most cases, we would not advocate for expanding an already burdensome series of laws, policies, and procedures. The singular intent of IRA is to provide a policy justification for rapid military support under imminently serious conditions when time does not permit commanders to obtain senior leader approval. Adding layers to and expanding the language of a policy intended to ensure speed and flexibility under dire circumstances seems counter-intuitive. However, the single most debated activity during the entire joint response to Hurricane Sandy occurred under the questionable justification of IRA, hence the motivation to suggest changes.

Many with direct knowledge and experience of the response in New York refute the justification offered by commanders that the Marines came ashore under IRA, noting that their arrival occurred 6 days after the storm's initial landfall and without urgent need or "imminently serious conditions."[19] Moreover, according to the same sources, the initial Marine activities on Staten Island did not "save lives, prevent human suffering, or mitigate great property damage."[20] Therefore, according to many, these actions do not constitute IRA. Others dispute this argument and reaffirm the Marines' support to the residents of Staten Island was justified under IRA, as they were requested to come ashore by a civil authority and in response to an immediate need as determined by the authorities on the ground.

One position remains consistent among those we spoke with, however. Regardless of the circumstances leading to or the justification for the Marines' support in New York, the activities carried out by the Marines post-Sandy were extremely beneficial to the residents and local authorities. So, while nobody debates the positive impact the Marines had on the Sandy response, the argument over IRA has been and will continue to be debated. Regardless of position, this debate centers on the subjective and often widely varied interpretation of the IRA policy in its current form. Our recommendations for changing the IRA policy address the primary concerns voiced by the majority of our data sources. At the very least, DoD should consider the preceding suggestions and form their own assessments by evaluating the utility and applicability of the content as it would be applied to a future DSC-led DSCA response.

Legislative and Associated Policy Revisions.

10 U.S.C. § 12304a.

Section 515 of the Fiscal Year 2012 National Defense Authorization Act added the legal authority for the Secretary of Defense to activate Reserve forces of the Army, Navy, Air Force, and Marine Corps in response to a governor's request for federal assistance during a disaster or emergency.[21] On December 31, 2011, 10

U.S.C. § 12304a became law and was implemented for the first time during the DoD response to Hurricane Sandy in New York with the activation of the three separate Army Reserve quartermaster detachments.[22] As with the larger DSC structure, the first attempt to implement this newly adopted statute resulted in some notable issues.

Sandy reports suggest that, although approved mission assignments were generated for the Army Reserve units in New York, coordinating with these detachments proved challenging. While the details of these challenges are vague at best, the recurrence of the issue across multiple sources suggests that DoD needs to improve Reserve component activation policies, and procedures under 10 U.S.C. § 12304a, in consonance with the recommendations noted in the Reserve Forces Policy Board's 2012 Information Memo on Reserve Component Operations in the Homeland.[23] Since the Reserve component is now a force-sourcing solution for DoD during disasters and emergencies, federal response capabilities and capacities are even greater. To maximize the effective use of the Reserve component during such incidents, each service branch must also implement policies detailing the activation procedures for their respective Reserve units under 12304a, including circumstances when Reserve units will be activated and under what capacity.

Establishing defined policies and procedures for reserve unit activation and sourcing under 12304a will improve an appointed DSC's ability to manage a joint operation. Due to its widespread geographic distribution throughout the continental United States, the Reserve component is a significant force multiplying asset that should be integrated into emergency and disaster response when required. This newly adopted legislation needs to be followed by service-specific policies that will ensure efficient mobilization and deployment of Reserve units in future DSC-led DSCA operations. Just as the Reserve component can now be a viable sourcing solution for Title 10 response efforts, so, too, can the National Guard.

32 U.S.C. § 502f.

Under 32 U.S.C. § 502f, the National Guard (or a member thereof) may "be ordered to perform training or other duty . . . (2) that may include . . . (A) support of operations or missions undertaken by the member's unit at the request of the President or Secretary of Defense."[24] When using 502f to activate the National Guard, troops remain under State control while support operations are funded 100 percent by DoD. Because states often are unable to fund their National Guard forces fully under State Active Duty for more than a few days, 502f provides a legal mechanism to relieve states of a funding dilemma. Historically, this legislation has been used as federal authority to mobilize the National Guard during nationally significant disaster responses such as Hurricane Katrina, as well as pre-planned national special security events.[25] As this statute offers states a mechanism to maintain control of the National Guard at 100 percent cost share to the Federal Government, it is clearly advantageous for states to request approval of a 502f designation during a presidentially declared disaster. However, due to ambiguity in the law, combined with states' desires for maximum control at minimum cost, states regularly request 502f designation from DoD. While some requests are approved, many are subsequently denied. Revising the current 502f language is necessary to address some of these issues, especially as it applies to force sourcing decisions within the DSC construct.

The main objective of the DSC during a no-notice/limited-notice incident like Sandy is to promote unity of effort between the National Guard and the Armed Forces. As such, the DSC should not be strategically or even operationally focused. The DSC should be a resource employer: a tactically focused commander looking to send the right resources to the right place at the right time. This general officer serves as a coordination mechanism between states and the Federal Government, and should not be concerned with the legal nuances and interpretations limiting National Guard duty statuses and funding source determinations. The DSC should possess a working knowledge of such information so as to appropriately influence tactical decisionmaking. However, National Guard duty status should be externally adjudicated to the DSC purview. If the DSC can use the National Guard to fill a request for assistance intended for Title 10 forces, he or she should not be limited in employing the necessary or available resources simply due to statutory nuances. Changing the current 502f wording to include specific criteria or guidelines for 502f designation and subsequent sourcing solutions can add strategic, operational, and tactical value to future DSCA operations.

Thus, DoD, the National Guard Bureau, and the states should establish specific criteria for 502f designation that includes the type(s) of incident(s) and/or circumstance(s) leading to a 502f authorization and under what circumstances a DSC can use 32 U.S.C. § 502f forces as a sourcing solution in place of Title 10 forces. Building on the recommendations of the Reserve Forces Policy Board's 2012 Information Memo noted earlier, these revisions should include criteria such as complex catastrophes, national significance or impact, or multistate response.[26] Defining such criteria and force-sourcing procedures will minimize the time required to allocate Title 32 resources if requested by the DSC. The 502f revisions, coupled with 12304a revisions, will address two notable gaps in the DSCA response to Sandy. While the DSC does not need to be an expert on the legal discussion, ignorance to the relevant laws is intolerable. As such, lawyers can and should be included as part of the joint staff in future DSC-led missions.

Inclusion of Staff Judge Advocate as part of Joint Task Force Headquarters Staff.

If you tell military commanders to cut through the red tape and make things happen, as was the case in Sandy, it is often the lawyer, or staff judge advocate (SJA), who is excised from the command decision-making process. With the numerous legal complexities and considerations that arise during a DSC-led DSCA response, excluding the SJA and overlooking laws and regulations leads to greater challenges during and after the incident. In some cases, the perceived urgency of a no-notice/limited-notice response effort and the need to provide assistance takes precedence over necessary legal considerations for managing and employing state and federal forces. Some decisions made during the Sandy response were of questionable legality and contrasted with the interpretations or advice of participating legal officers.[27] In other instances, SJAs were not provided an opportunity to advise commanders prior to such decisions.[28] As Sandy was the first attempt at using the DSC arrangement in this capacity, these issues are expected but should be addressed for future operations.

However cumbersome, nuanced, and seemingly arbitrary these laws may seem to commanders focused on accomplishing a mission, laws are written to provide structure and limits. Within the context of the DSC, many of the relevant laws are rooted in the Constitution and the foundational principles by which we govern our Armed Forces. Lawyers arguably provide some of the most critical knowledge during a combined state and federal military response; they cannot be excluded from advising the DSC on the statutory limitations of military actions under unpredictable circumstances.

As part of its effort to develop a DSC instruction for DSCA operations, DoD should incorporate policy guidance that encourages the use and active employment of DSCA-knowledgeable SJA personnel as contributing members of future JTF staffs. Consideration should also be given to employing two attorneys: one with Title 10 knowledge and oversight and one with Title 32 knowledge and oversight. Including SJAs in future DSCA staffs will enhance the operational and legal integrity and minimize future issues like those encountered following Sandy. However, legal knowledge alone is not sufficient to improving future DSCA operations under the DSC construct. The confusion among Title 10 forces during this response points to a critical need to improve DSCA education in future Title 10 officers.

Expand and Reinforce DSCA Education for Officers.

If Sandy is a barometer for the state of DSCA knowledge among Title 10 commanders, there is significant room for improvement. Not only were some Title 10 commanders unaware of who the DSC was or how to contact him and his staff, but also some officers had never heard of the DSC construct prior to Hurricane Sandy.[29] Active Component forces demonstrated a degree of ignorance or disregard to the mission assignment process that was reaffirmed through command guidance. By abandoning processes and procedures, some Title 10 forces supplanted (rather than supported) local authorities' efforts. Likewise, U.S. Army Corps of Engineers (USACE) personnel were equally unfamiliar with the DSC construct and the statutory limitations over Title 10 forces in support of ESF 3. While not a blanket indictment of the military officer corps or the USACE, as only a small sample participated in Sandy, these issues are just some of the many encountered during the DSCA response, further suggesting that improvements to DSCA education are necessary.

Most, if not all, top-level DoD schools offer some degree of DSCA education through practical application exercises, classroom instruction, or a combination of both. Many officers also receive in-depth instruction on or exposure to the topic through the completion of these and other capstone-type projects pursued while in residence. So, while most commanders have received at least some exposure to DSCA, the inherent complexities and fluidity of the DSCA environment require constant refresher training. Not all commands/billets require the same level of DSCA knowledge, however. Therefore, DoD, with the support of the individual services, should identify and designate DSCA-relevant command billets required to complete annual DSCA training. Following an assessment of expected DSCA requirements, capabilities assessments should help identify DSCA-capable units and their respective command billets. An example of a DSCA-capable unit is the Marine Expeditionary Unit (MEU). MEU commanders from I and II MEF (Marine Expeditionary Force, continental United States) should receive annual refresher training similar in format to the currently offered DSCA courses via Joint Knowledge Online.

In addition to identifying DSCA-relevant command billets and requiring refresher training, DSCA education should occur during basic officer training and continue during subsequent professional military education and career level schools. Marine officers attending The Basic School, for example, should receive DSCA familiarization training via classroom instruction during Phase IV of the course curriculum. Following initial exposure in entry level schools, officers will have a foundational understanding of the subject to leverage as they progress through future professional military education. Adopting such educational requirements for company grade officers will ensure those officers slated for top-level school and future command billets at DSCA-capable units possess the necessary and continued education to facilitate operationally and tactically sound decisionmaking in future DSCA environments.

ENDNOTES - CHAPTER 5

1. Interviews conducted by the author with various National Guard officers and active military officers that participated in the response to Hurricane Sandy in some capacity, January-March 2014.

2. Interviews conducted by the author with various DoD officials with relevant knowledge of the circumstances before, during, and after the Sandy response, March-May 2014.

3. Department of Defense, *Joint Publication (JP) 3-28, Defense Support of Civil Authorities,* Washington, DC: U.S. Government Printing Office, July 31, 2013, p. II-12.

4. Interviews conducted by the author with various DoD employees, National Guard officers, and active military officers.

5. JP 3-28, p. C-1.

6. Federal/state cost share/reimbursement considerations are beyond the scope of the DSC. If this recommendation were to be implemented in the future, revisions to relevant legislation (Stafford Act, etc.) should be considered, with possible changes that would include a period of 100 percent federal cost share for the duration of the proposed Title 10 integration period.

7. *DoD Directive 3025.18, Defense Support of Civil Authorities,* Washington, DC: DoD, p. 4.

8. Department of Defense, *Joint Publication 3-03, Joint Operations,* Washington, DC: U.S. Government Printing Office, August 11, 2011, pp. xi, I-8.

9. JP 3-28, p. C-1.

10. Interviews conducted by the author with multiple DoD officials with relevant knowledge of DSCA policies, May 2014.

11. Interview conducted by the author with a National Guard officer that participated in the response to Hurricane Sandy, February 2014.

12. DoD Directive 3025.18, p. 4

13. Interviews conducted by the author with multiple DoD officials with relevant knowledge of DSCA policies, May 2014.

14. This claim is supported through interviews with multiple officers of both the National Guard and Active Component who participated in the Sandy response effort.

15. Department of Defense, *JP 1-02: Department of Defense Dictionary of Military and Associated Terms*, Washington, DC: U.S. Government Printing Office, March 15, 2014, p. 36.

16. DoD Directive 3025.18, p. 4.

17. Dunphy and Radel (2009) offer a useful approach to evaluating this process. Their study methodology can be duplicated with emphasis on the mission assignment process. See Charles Dunphy, Jr., and Christopher Radel, "Assignment of Local, State, and Federal Request Processes for Defense Support to Civil Authorities in the United States," Monterey, CA: Naval Postgraduate School, December 2009.

18. See Marine Corps Fiscal Year 2012 recruiting campaign, "Towards the Sounds of Chaos." In order to appeal to a growing sense of altruism and public service in the millennial generation, the Marine Corps adopted the aforementioned campaign with commercials depicting Marines engaged in humanitarian assistance/disaster response missions, among other things. Author's personal knowledge of and experience with Marine Corps recruiting initiatives.

19. Interviews conducted by the author with various DoD employees, National Guard officers, and Active Component officers involved in the Sandy response.

20. *Ibid.*; DoD Directive 3025.18, p. 4.

21. National Defense Authorization Act for Fiscal Year 2012, Public Law 112-81, H.R. 1540, p. 98.

22. Office of the Secretary of Defense After Action Review-Hurricane Sandy, January 10, 2013, p. 32.

23. *Report of Reserve Forces Policy Board on New Policies and Clearer Funding Flows for Reserve Component Operations in the Homeland,* Falls Church, VA: Department of Defense, April 9, 2012, p. 7, available from *rfpb.defense.gov/Portals/67/Documents/RFPB%20Memo%20to%20SECDEF-Policy%20and%20Funding%20RC%20Homeland-9%20Apr%202012.pdf.*

24. 32 U.S.C. § 502, "Required Drills and Field Exercises," Subsection (f). 1964, 2006.

25. Interviews conducted by the author with various DoD employees and National Guard officers who participated in Hurricane Sandy.

26. *Report of Reserve Forces Policy Board on New Policies and Clearer Funding Flows for Reserve Component Operations in the Homeland,* p. 10.

27. Interviews conducted by the author with three military judge advocates, including one who served as an on-the-ground observer of Title 10 activities during the Sandy response in New York, December 2013-January 2014. These assertions are corroborated across multiple DoD after-action reports reviewed by the authors while conducting this research.

28. *Ibid.*

29. Interviews conducted by the author with various DoD employees, National Guard officers, and Active Component officers involved in the Sandy response, January-March 2014, and reaffirmed through document analysis of after-action reports and other Sandy-relevant resources.

CHAPTER 6

CONCLUSION

The U.S. military's primary mission is to fight and win our nation's wars. In this regard, our nation's military will continue training for combat operations and other contingency missions around the world. However, with the ongoing defense drawdown from combat operations in Afghanistan, the military will now look to enhance its civil support readiness as a priority domestic mission focus. The reality is that, when a large-scale event occurs, the Department of Defense (DoD) and its assets can provide timely and extensive support beyond the capacity of any state or local government agency. Combining Title 10 and Title 32 forces only further multiplies this already unparalleled capability. Written into law as the "usual and customary" arrangement during the simultaneous employment of the National Guard and Armed Forces,[1] the dual status commander (DSC) arrangement serves as the coordination mechanism that should enable the efficient and effective integration and employment of military forces to meet the needs of those affected by future disasters and emergencies. With the events of Hurricane Sandy behind us, now is the time to learn from this historic response and determine ways to improve future military civil support efforts under the DSC construct.

The preceding analysis offered suggestions aimed at improving the mechanics of the DSC process through various operational strategy and policy-oriented recommendations. The DSC concept shows promise and has been used again in more recent events with notable success.[2] While the concept is sound, the execution during Sandy was flawed. Failing to acknowledge and improve upon the lessons learned from Sandy will question the efficacy of using DSCs for future response efforts. If we truly want to commit to the DSC as the usual and customary command arrangement as the law states, we need to maximize the use of this and other analyses to repeat the successes and avoid the failures in future operations.

With such a large and complex mission to coordinate, including the consideration of relevant laws, procedures, and command authorities, many of which are embedded in our Constitution, the DSC construct provides the necessary structure to facilitate effective Defense Support of Civilian Authorities operations between States and the Federal Government. While improvisation, adaptability, and flexibility are valued aspects of military operational doctrine and mission-oriented command and control, the complexities of no-notice/limited-notice response missions require some semblance of organization and boundaries.

Hurricane Sandy was a significant event; but it was not a catastrophe. The urgency of the federal response and the lack of adherence to policies and procedures added to the confusion in some cases. We cannot forecast future requirements, nor can we predict how future operations will transpire. There will always be a level of uncertainty and a sense of urgency during no-notice/limited-notice incidents. We can, however, mitigate, in part, future uncertainty and confusion through the application of lessons learned, such as those provided in this analysis. By identifying and incorporating lessons learned into future incident response, we can continue our efforts to mature these complex operations. Such improvements will likely lead to increased capability of military personnel; enhanced knowledge for those unfamiliar with the identified concepts; and ultimately, more lives, property, and resources saved in the aftermath of the next event requiring defense support of civil authorities.

ENDNOTES - CHAPTER 6

1. National Defense Authorization Act for Fiscal Year 2012, Public Law 112-81, H.R. 1540, p. 98.

2. A DSC was appointed and commanded Joint Task Force Centennial, a combined Title 10 and 32 operation, during the September 2013 response to the Colorado Floods.

APPENDIX I

ACRONYMS

AAR	After Action Report
AC	Active Component
ADP	Army Doctrinal Publication
ARG	Amphibious Ready Group
ARNORTH	Army North
BG	Brigadier General
CBRN	Chemical, Biological, Radiological, Nuclear
CG	Commanding General
CJCS	Chairman of the Joint Chiefs of Staff
CJCSG	Chairman of the Joint Chiefs of Staff Guidance
CJCSM	Chairman of the Joint Chiefs of Staff Manual
COCOM	Combatant Command
CONUS	Continental United States
COP	Common Operating Picture
DCO	Defense Coordinating Officer
DLA	Defense Logistics Agency
DOD	Department of Defense
DODD	Department of Defense Directive
DSCA	Defense Support of Civil Authorities
DSC	Dual Status Commander

DSC JTF	Dual Status Commander Joint Task Force
EMAC	Emergency Management Assistance Compact
EPLO	Emergency Preparedness Liaison Officer
ESF	Emergency Support Function
EXORD	Execution Order
FCO	Federal Coordinating Officer
FEMA	Federal Emergency Management Agency
FM	Field Manual
FRAGO	Fragmentary Order
GAO	Government Accountability Office
GOV	Governor
HD	Homeland Defense
HD/ASA	Homeland Defense and America's Security Affairs
HQMC	Headquarters Marine Corps
HS	Homeland Security
IRA	Immediate Response Authority
JBMDL	Joint Base McGuire Dix Lakehurst
JCE	Joint Coordinating Element
JFHQ	Joint Force Headquarters
JFLCC	Joint Force Land Component Commander
JFMCC	Joint Force Maritime Component Commander

JKO	Joint Knowledge Online
JOA	Joint Operations Area
JP	Joint Publication
JRSOI	Joint Reception Staging Onward Movement and Integration
JTF	Joint Task Force
JTF-CS	Joint Task Force Civil Support
LNO	Liaison Officer
LTG	Lieutenant General
MA	Mission Assignment
MA	Massachusetts
MD	Maryland
MEF	Marine Expeditionary Force
MEU	Marine Expeditionary Unit
MG	Major General
MOA	Memorandum of Agreement
MTTP	Multi-Service Tactics Techniques and Procedures
NDAA Act	National Defense Authorization
NG	National Guard
NGB	National Guard Bureau
NH	New Hampshire
NIMS	National Incident Management System
NJ	New Jersey
NRF	National Response Framework
NSSE	National Security Special Event

NORTHCOM	U.S. Northern Command
NY	New York
NYC	New York City
OEM	Office of Emergency Management
OSD	Office of the Secretary of Defense
PA	Pennsylvania
PCA	Posse Comitatus Act
PL	Public Law
PME	Professional Military Education
POTUS	President of the United States
PTDO	Prepare to Deploy Orders
RFA	Request for Assistance
RI	Rhode Island
SAD	State Active Duty
SCO	State Coordinating Officer
SECDEF	Secretary of Defense
SJA	Staff Judge Advocate
T10	Title 10
T32	Title 32
TACON	Tactical Control
TAG	The Adjutants General
TF	Task Force
TRANSCOM	United States Transportation Command
TS	Tropical Storm
UOC	Unity of Command
UOE	Unity of Effort

US	United States
USA	United States Army
USACE	United States Army Corps of Engineers
USAF	United States Air Force
USAR	United States Army Reserve
USC	United States Code
USCG	United States Coast Guard
USFF	United States Fleet Forces
USMC	United States Marine Corps
USN	United States Navy
USS	United States Ship
VOCO	Verbal Orders of the Commanding Officer

APPENDIX II

DUAL STATUS COMMANDER DESIGNATION PROCESS

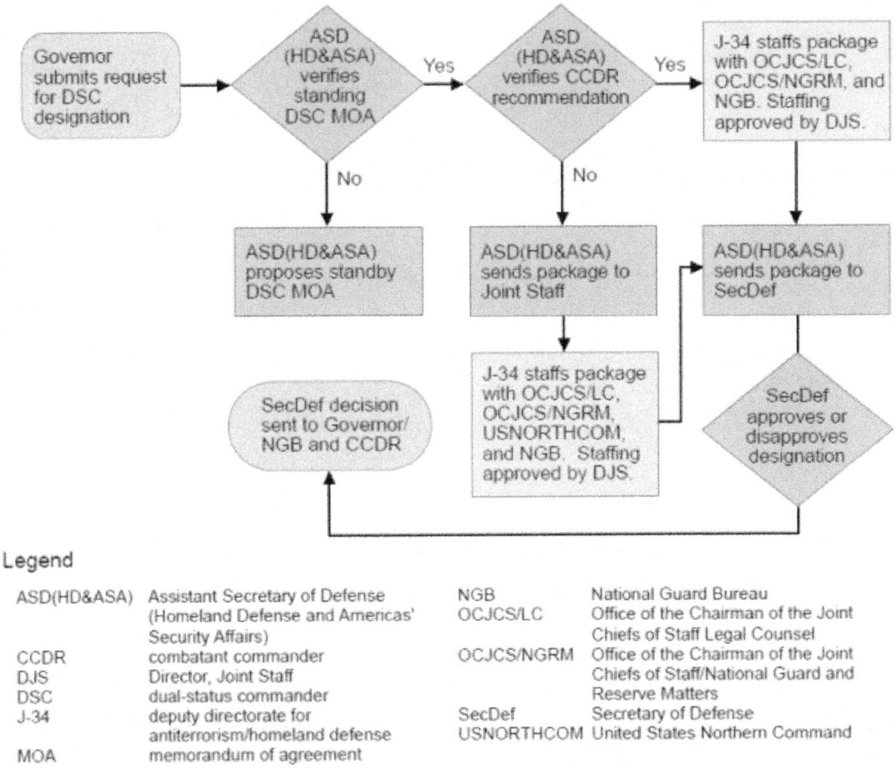

Source: Department of Defense, *Joint Publication 3-28, Defense Support of Civil Authorities*, Washington, DC: U.S. Government Printing Office, 2013, p. C-9.

Figure II-1. DSC Designation Process.

U.S. ARMY WAR COLLEGE

Major General William E. Rapp
Commandant

*****

STRATEGIC STUDIES INSTITUTE
and
U.S. ARMY WAR COLLEGE PRESS

Director
Professor Douglas C. Lovelace, Jr.

Director of Research
Dr. Steven K. Metz

Author
Mr. Ryan Burke
Dr. Sue McNeil

Editor for Production
Dr. James G. Pierce

Publications Assistant
Ms. Rita A. Rummel

*****

Composition
Mrs. Jennifer E. Nevil

Made in the USA
Monee, IL
04 November 2026

67384045R00079